Springer Theses

Recognizing Outstanding Ph.D. Research

Aims and Scope

The series "Springer Theses" brings together a selection of the very best Ph.D. theses from around the world and across the physical sciences. Nominated and endorsed by two recognized specialists, each published volume has been selected for its scientific excellence and the high impact of its contents for the pertinent field of research. For greater accessibility to non-specialists, the published versions include an extended introduction, as well as a foreword by the student's supervisor explaining the special relevance of the work for the field. As a whole, the series will provide a valuable resource both for newcomers to the research fields described, and for other scientists seeking detailed background information on special questions. Finally, it provides an accredited documentation of the valuable contributions made by today's younger generation of scientists.

Theses are accepted into the series by invited nomination only and must fulfill all of the following criteria

- They must be written in good English.
- The topic should fall within the confines of Chemistry, Physics, Earth Sciences, Engineering and related interdisciplinary fields such as Materials, Nanoscience, Chemical Engineering, Complex Systems and Biophysics.
- The work reported in the thesis must represent a significant scientific advance.
- If the thesis includes previously published material, permission to reproduce this must be gained from the respective copyright holder.
- They must have been examined and passed during the 12 months prior to nomination.
- Each thesis should include a foreword by the supervisor outlining the significance of its content.
- The theses should have a clearly defined structure including an introduction accessible to scientists not expert in that particular field.

More information about this series at http://www.springer.com/series/8790

Abhijit Saha

Molecular Recognition of DNA Double Helix

Gene Regulation and Photochemistry of ^{Br}U-Substituted DNA

Development of Chemical Biology Tools

Doctoral Thesis accepted by
Kyoto University, Kyoto, Japan

Springer

Author
Dr. Abhijit Saha
Institute Curie
Orsay
France

Supervisor
Prof. Hiroshi Sugiyama
Department of Chemistry, Graduate School of Science, Institute for Integrated Cell-Material Sciences (WPI-iCeMS)
Kyoto University
Kyoto
Japan

ISSN 2190-5053 ISSN 2190-5061 (electronic)
Springer Theses
ISBN 978-981-10-8745-5 ISBN 978-981-10-8746-2 (eBook)
https://doi.org/10.1007/978-981-10-8746-2

Library of Congress Control Number: 2018934869

Printed on acid-free paper

This Springer imprint is published by the registered company Springer Nature Singapore Pte Ltd.
part of Springer Nature
The registered company address is: 152 Beach Road, #21-01/04 Gateway East, Singapore 189721, Singapore

This book is lovingly and sincerely dedicated to my father ***Late Akhil Chandra Saha*** *who was a great source of inspiration and motivation*

Supervisor's Foreword

During the last three decades, we have seen enormous development in the field of DNA molecular recognition. There have been plenty of small molecules developed to target specific DNA sequences in the human genome. Small molecule targeted therapeutics is increasingly a choice for anticancer drugs and many other diseases. Our group contributed for last two decades in developing small molecules for therapeutic applications and this thesis demonstrates some developments in this area.

In this thesis, Dr. Abhijit Saha made an effort to design and synthesize small molecules to target specific HDAC inhibition. Though small molecules for HDAC inhibition are known for several years they are nonspecific in nature. Thus, sequence-specific HDAC inhibition by small molecules could contribute in understanding the complex transcription network. The research described in Chaps. 2 and 3 is the development of a small molecule which can express pluripotent genes and developmental genes from mouse embryonic fibroblast by selective HDAC inhibition.

We also investigated the photochemistry of ^{Br}U-substituted DNA for the last three decades in order to understand the underlying mechanism of cleavage. In this thesis, some useful applications of this modified DNA are shown. In the design and development of small molecules, it is always necessary to screen a library of potential binding sites to identify real highest affinity binding sites for the application in biology. The author developed a photo-footprinting method for the detection of the binding sites of pyrene-conjugated polyamides using ^{Br}U-substituted DNA. This method can be useful on a routine basis for detecting the binding sites of small molecules.

^{Br}U-substituted DNA was further exploited in the detection of cooperative binding of transcription factors, Sox2 and Pax6, on a ^{Br}U-labeled regulatory element DC5 by excess electron transfer. With this tool, it is possible to understand complex protein–nucleic acids interactions during their cooperative binding and it may be a useful tool for other biological complex interactions.

At final part of the thesis, the author evaluated the effect of Hoechst 33258 dye in the double-strand break of ^{Br}U-labeled DNA after UVA irradiation.

This thesis demonstrates the development of small molecule for the applications in regenerative medicines. Also, it demonstrates the applications of ^{Br}U-based tool to study the binding of DNA-binding ligands and proteins.

Kyoto, Japan
December 2017

Prof. Hiroshi Sugiyama

Parts of this thesis have been published in the following journal articles:

1. A. Saha, Ganesh N. Pandian, S. Sato, J. Taniguchi, K. Hashiya, T. Bando and H. Sugiyama (2013) "Synthesis and biological evaluation of a targeted DNA-binding transcriptional activator with HDAC8 inhibitory activity" *Bioorg. Med. Chem.* 21, 4201–4209. (doi: 10.1016/j.bmc.2013.05.002)
2. A. Saha, G. N. Pandian, S. Sato, J. Taniguchi, Y. Kawamoto, K. Hashiya, T. Bando, and H. Sugiyama, (2014) "Chemical Modification of a Synthetic Small Molecule Boosts its Biological Efficacy against Pluripotency Genes in Mouse Fibroblast" *ChemMedChem,* 9, 2374 (doi: 10.1002/cmdc.201402117)
3. A. Saha, F. Hashiya, S. Kizaki and H. Sugiyama (2015) "A novel detection technique of polyamide binding sites by photo-induced electron transfer in ^{Br}U substituted DNA" *Chem. Commun. 45*, 14485–14488 (doi:10.1039/C5CC05104E)
4. Abhijit Saha, Seiichiro Kizaki, Deboyoti De, Masayuki Endo, Kyeong Kyu Kim, Hiroshi Sugiyama (2016) "Examining cooperative binding of Sox2 on DC5 regulatory element upon complex formation with Pax6 through excess electron transfer assay" *Nucleic Acids Research*, 44, e125 (doi: 10.1093/nar/gkw478)
5. A. Saha, S. Kizaki, Ji Hoon Han, Zutao Yu H. Sugiyama (2018) "UVA irradiation of ^{Br}U-substituted DNA in the presence of Hoechst 33258" *Bioorganic & Medicinal Chemistry*, 26, 37–40 (doi: 10.1016/j.bmc.2017.11.011)

Acknowledgements

First and foremost, I express my sincere gratitude to **Prof. Hiroshi Sugiyama** for his guidance, valuable discussions, support, and inspirations throughout my Ph.D.

I am also deeply grateful to Dr. Toshikazu Bando (Associate Professor, Kyoto University), and Dr. Ganesh N. Pandian (Assistant Professor, iCEMs) for their discussions, which improved my course of study substantially. I also thank Dr. M. Endo and Dr. Soyeong Park for their guidance and support.

I am grateful to some great teachers in my life during my undergraduate studies. They are also responsible for my success. I express my sincere gratitude to them especially Dr. Sanjib Deuri, Late Shanti Ranjan Biswas, Mrs. Dipanjali Pathak, Mr. B. R. Chakroborty, Mr. Manmohan Nath, Dr. Malay Kumar Barman, and Ramen Das.

I extend my gratitude to Dr. Bhubaneswar Mandal, (Associate Professor, Indian Institute of Technology Guwahati, India) and Prof. Biprajit Sarkar (Professor, Berlin University, Germany) for their support.

I thank all my friends especially Himashree Pathak, N. K. Chaitanya, Ashim Paul, Thalluri Kishore, Partha Dey, Bhaskar Nath, Somnath Bhowmick, Rhys D. Taylor, Seiichiro Kizaki, Anandhkumar Chandran, Junetha Syed, Yue Li, and S. Sato for their support and help.

I am highly gratified with the care and help from Yasuko Niimi (Secretary) and Kaori Hashiya (Technician) during my stay at Kyoto.

I sincerely believe that without the kind help from my other lab mates, the work would not have been possible. I thank them for their support.

Finally, I thank all my family members for their full support and inspirations during my Ph.D.

Last but not the least; I acknowledge the support from Seiwa International Scholarship Foundation for partial support in my doctoral program.

Contents

Chapter 1
Overview of DNA Minor Groove-Binding Synthetic Small Molecules and Photochemistry of ^{Br}U-Substituted DNA

Abstract DNA was the first defined target for the anticancer drugs. Targeting DNA with small molecules for therapeutic applications is in increasing demand. Pyrrole–imidazole polyamide is one such molecule, which has shown promising selectivity for specific DNA sequences of the genome. This can bind to the minor groove DNA in a sequence-specific manner and the binding efficiency is comparable to the natural transcription factors. This molecule was used extensively to study gene regulations. In the first part of this thesis, development of this small molecule for specific gene activation is demonstrated for the purpose of regenerative medicine. In the second part, the photochemistry of ^{Br}U-labeled DNA is used to develop ^{Br}U-based detection assay. Several applications of this photochemistry are demonstrated such as the development of a photo-footprinting technique to identify the binding sites of pyrrole–imidazole polyamides. This photochemistry is also used for detection of cooperative binding of transcription factors on ^{Br}U-labeled regulatory element and the double-strand breaks in ^{Br}U-labeled DNA by Hoechst 33258. The photochemical methods described in this thesis are useful for studying small molecules and proteins binding on DNA.

Keywords Molecular recognition · Pyrrole–imidazole polyamide
Transcriptional activator · 5-Bromouracil · UV irradiation · Photochemistry

1.1 General Introduction: (A) Molecular Recognition of DNA

Living organisms store their genetic information in the DNA double helix. The genomic DNA is a polymer of A, T, G, and C and their unique sequences all over the genome are crucial for life. In this new era of genetics, it is possible to find the link between human diseases and the crucial changes in the DNA sequences. Besides this genetic encoding, the DNA has several conformations such as B-form, A-form, Z-form, cruciform, triplex DNA and four-stranded G-quadruplex (Fig. 1.1). In B-form, the most common double helical structure found in nature, the double helix

A. Saha, *Molecular Recognition of DNA Double Helix*, Springer Theses,
https://doi.org/10.1007/978-981-10-8746-2_1

is right handed with 10–10.5 base per turn. The double helix in it contains a wider major groove and a narrow minor groove. In A-DNA, the conformation adopted by DNA–RNA hybrids, contains a deep and narrow major groove. In Z-DNA, the double helix is left-handed and the conformation is favored for alternating G-C sequences. Cruciform structure forms by negative supercoiling of a B−DNA to adopt a four-armed, cruciform secondary structure that resembles a Holliday junction. These structures require ≥ 6−nucleotide inverted repeats (cruciform motif) to form, and such motifs are located near replication origins, breakpoint junctions and promoters in diverse organisms. Triplex DNA forms when a single-stranded DNA forms Hoogsteen hydrogen bonds in the major groove of purine-rich double-stranded B-DNA. Triplexes can form at physiological pH, and these structures are also stabilized by negative supercoiling. G-quadruplexes are four-stranded DNA structures formed by G-rich sequence [1]. Such local DNA conformations have been suggested to be biologically important in processes such as DNA replication, gene expression, and regulation, and the repair of DNA damage [2].

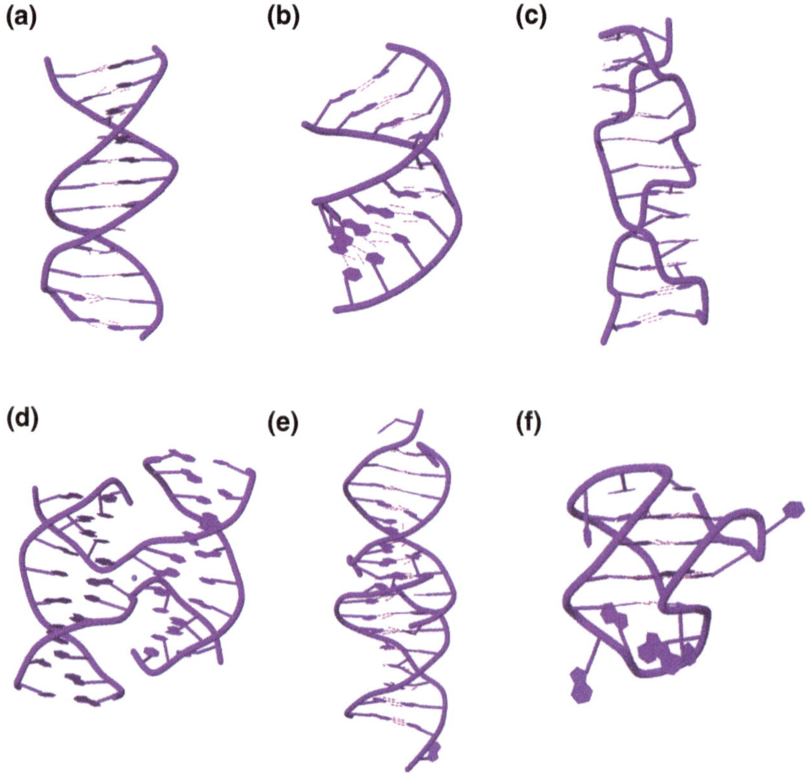

Fig. 1.1 Structures of **a** B-DNA, **b** A-DNA, **c** Z-DNA, **d** cruciform, **e** Triplex-DNA, and **f** G-Quadruplex DNA

In nature, proteins can read out predetermined DNA sequences and control the biological functions. Sequence-specific interactions between proteins and DNA are chemically complex [3]. Crystal structures of protein DNA complexes have solved many of such complicated interactions [4]. Proteins are relatively large in size and thus can fit predominantly into the major groove (groove width and depth 11.7 and 8.8 Å). Well-known proteins including helix turn helix [5], zinc finger motif [6], homeobox domain [7], and bZip motif [8] are known as major groove binder. Nature indeed provides clues to build organic small molecules/natural products, which can distinguish each of the four Watson–Crick base pairings. DNA-binding organic small molecules, natural products can preferentially fit into the minor groove because of its narrow width and depth 5.7 and 7.5 Å respectively. For several years, DNA minor groove in the B-DNA is of great interest for developing new drugs since it is a site of non-covalent high sequence-specific interactions for a large variety of organic small molecules including natural products. DNA minor groove-binding small molecules have been extensively studied because of their several biological properties. In fact, they possess antibacterial, antiviral, antiprotozoal, and antitumor properties [9]. Apart from B-DNA, recently small molecules also developed to target G-quadruplexes as a potential target for cancer treatment. With the growing advancement of experimental techniques suitable for the determination of the DNA-binding specificity, for example, footprinting, affinity cleavage, X-ray crystallography, NMR, molecular modeling, surface plasmon resonance (SPR), and more recently introduced Bind-n-Seq have advanced the understanding of DNA–small molecule interactions and their nature of reactions [3, 10]. The knowledge of these various techniques permitted the synthesis of sequence-selective targets with better biological activities both in vitro and in vivo. To this context, pyrrole–imidazole polyamides achieved considerable interest as sequence-specific DNA minor groove binder. This small molecule has been used to target human genome based on its specific-genomic recognition and have evidence of right biological activity. Overcoming the new synthetic challenges, it is now possible to convert this molecule into a multifunctional molecule.

1.1.1 Molecular Recognition of Pyrrole–Imidazole Polyamides (PIPs) on B-DNA

At the early stage, natural products such as Distamycin A, Netropsin, chromomycin, Actinomycin D, calicheamicin oligosaccharide were characterized to bind selected DNA sequences [11]. Interestingly, some small molecules such as Hoechst 33,258, DAPI, berenil, pentamidine also have been developed which can bind minor groove of DNA [9]. Most of the natural products actinomycin, daunomycin, echinomycin, and chromomycin are structurally complex. Among them, the compounds having the simplest structures are Netropsin and Distamycin A with two and three consecutive pyrrole rings connected by amide

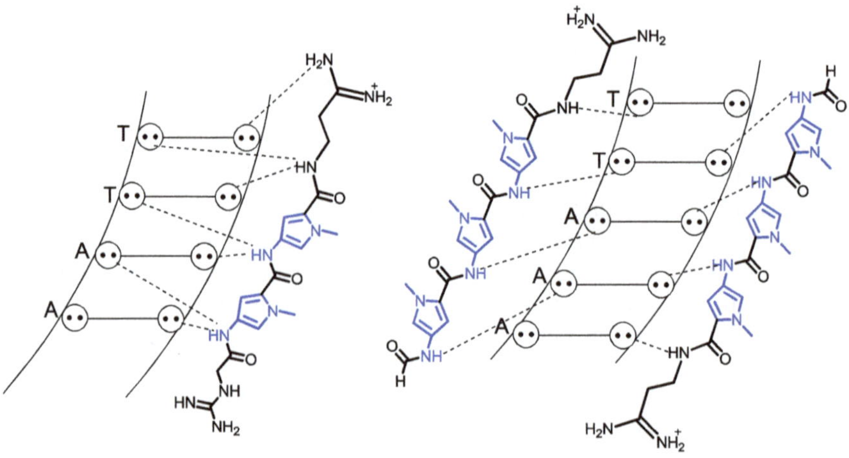

Fig. 1.2 Molecular recognitions of natural products distamycin and netropsin

bonds respectively (Fig. 1.2). These two compounds were studied very carefully to understand the molecular recognition of the DNA double helix. In this regard, the first X-ray structure of a complex of netropsin and DNA revealed that the 1:1 complex was a shape-selective recognition with the crescent PyPy bound by the walls of the narrow minor groove of consecutive A•T tract in the minor groove [12]. From the crystallographic analysis, it was found that NHs of the carboxamides pointed toward the minor groove floor of the helix making specific hydrogen bonds with the A•T and T•A base pairs (N3 of A and O_2 of T). Subsequently, another complex of distamycin (PyPyPy) and nucleic acids were solved by NMR and demonstrated that distamycin could bind A•T sequences of DNA as an antiparallel 2:1 complex as well as 1:1 [13].

Together this recognition of netropsin and distamycin into the nucleic acids led the foundation to design molecules with better recognition ability. It was then thought to replace pyrrole (Py) with imidazole (Im) rings in the polyamide which might allow to read the exocyclic NH_2 of G•C base pairs in the 1:1 complex [12, 14]. The synthesis of a novel polyamide, ImPyPy was reported which was expected to bind, according to the 1:1 model, the sequence 5′-(G,C)(A,T)$_2$-3′ in a single orientation [15]. After scanning a library of sites on several DNA restriction fragments using MPE-Fe(II) footprinting and affinity cleavage techniques, it was found that the molecule bound not the expected sequence but rather a new unanticipated five base pair sequence, 5-(W)G(W)C(W)-3 (where W=A/T). Surprisingly, G was located in the second position and C was preferred in the fourth position. This time the previous breakthrough NMR of distamycin in 2:1 complex led to the fact that polyamide ImPyPy binds as an antiparallel dimer, suggesting that pairs of Im/Py recognize G•C, Py/Im recognizes C•G but they do not recognize

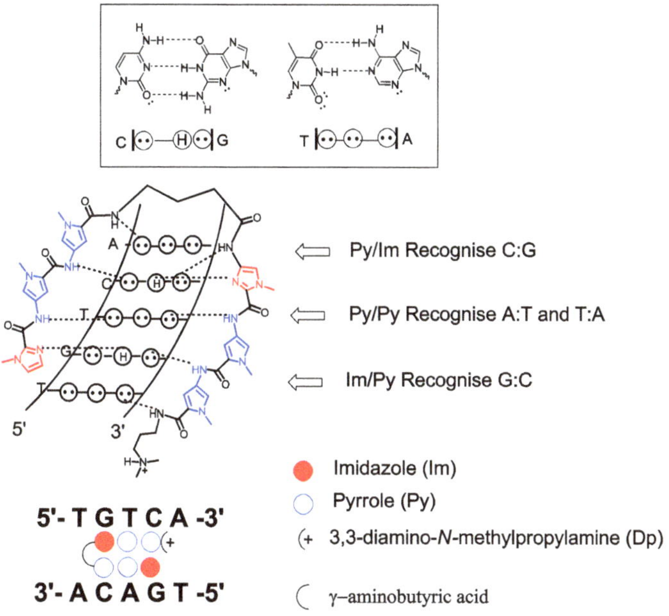

Fig. 1.3 Molecular recognition of DNA minor groove by a hairpin polyamide

either A•T or T•A [16]. The Py/Py pair cannot discriminate between T•A and A•T base pairs. In the next round of studies, these antiparallel dimers were covalently attached with an aliphatic amino acid (γ-aminobutyric acid) to create a U-shaped motif which bound the minor groove of DNA with very high affinity and specificity [17]. The hairpin structure kept the rings unambiguously "paired" avoiding slipped dimers. In Fig. 1.3, the hairpin polyamide originated from ImPyPy dimers by connecting the amino and carboxy terminus with γ-aminobutyric acid is shown.

1.1.2 Molecular Recognition by Small Molecules on Non-B-Form DNA

Like B-form DNA, G-quadruplex is also can be targeted by small molecules. However, the recognition of G-quadruplex by the small molecule is different from the recognition of PIPs on B-DNA. The reason is obvious from the structural point of view. Four-stranded G-quadruplex (G4) structures, composed of several layers of guanine quartets which are stabilized by Hoogsteen hydrogen bonds and coordination to central metal cations. This secondary structure has a planar surface on the top and bottom by the quartets. Thus, relying on the planar surface of the quartets, small molecules were designed to target this structure. Some examples of G-quadruplex binding small molecules are PhenDC3 [18] and Pyridostatin [19] (Fig. 1.4).

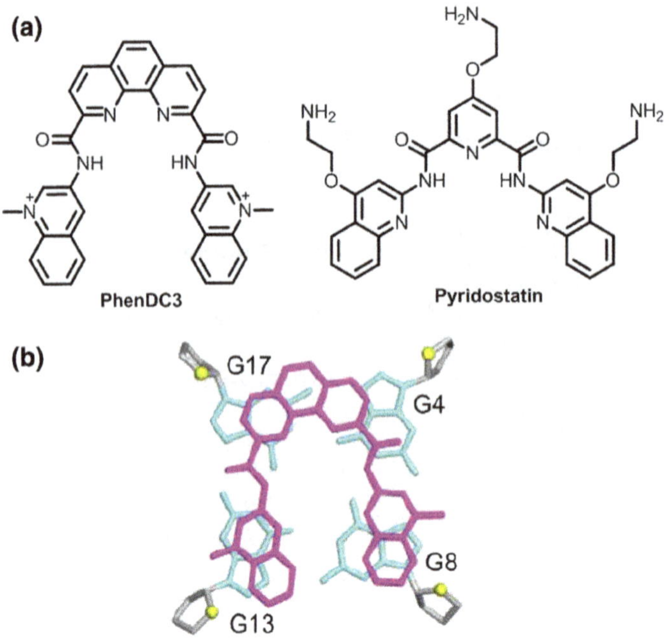

Fig. 1.4 **a** The chemical structure of G-quadruplex-binding ligands. **b** The NMR solution structure of the complex of PhenDC3 and G-quadruplex DNA

The binding of this class of ligands on this secondary structure is demonstrated by NMR solution structure of the complex of PhenDC3 and intramolecular G-quadruplex-derived c-myc sequence as shown in Fig. 1.4b. The ligand PhenDC3 interacts strongly through π-stacking with the G bases on the G-tetrads [20]. These compounds have shown promising biological activities in vivo.

1.1.3 Structural Modifications on PIPs for Better Properties

Later on, important chemical modifications were carried out such as introducing (a) chiral turn: employing a chiral turn on the γ-turn residue by amino-substitution at the α-position to achieve enhanced binding affinity (10-fold) without loss of specificity, higher orientational selectivity [21] (b) cyclic polyamide: cyclic polyamides show higher affinity with respect to the hairpin structure [22]. (c) ***β*-alanine/** ring pair: the ***β***/ring pair relaxes the ligand curvature and allows the hairpin structure to adjust to the microstructure of non-B-form helix. In some cases, the binding affinity of the ***β*-alanine**/ring-polyamides is significantly higher than that of the ring/ring analog [23]. (d) Tandem hairpin: the linked hairpins not only increase the binding site size also significantly increases binding affinity and specificity of the ligand (Fig. 1.5) [24].

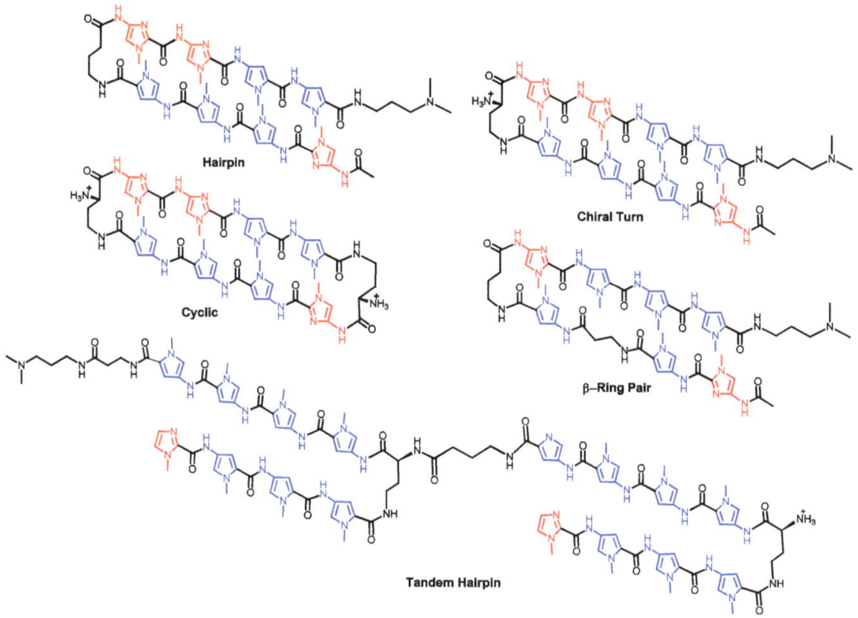

Fig. 1.5 Examples of different types of polyamide structures starting from hairpin, hairpin with a chiral turn, cyclic polyamide, *β*-alanine/ring hairpin, and tandem polyamides

1.2 Biological Activity

There are approximately 19,000–20,000 genes in each human cell. It is fascinating to realize that how nature controls the expression in time and space of each of these genes by a remarkable three-dimensional switch, that multiprotein complex assembled on specific base pairs of "promoter" DNA sequence encoded upstream from the RNA polymerase start site and coding region. Transcription factors bind very specific DNA sequences in the promoter region of each gene modulating the expression of that gene. Therefore, sequence-specific pyrrole–imidazole polyamide is a good candidate to compete with transcription factors and interferes the gene expression.

1.2.1 PIPs as Gene Down Regulator

PIPs have been extensively studied to inhibit specific gene of interest for many years. The binding affinities of polyamides are sufficient enough to inhibit and compete with transcription factors in order to downregulate a specific gene. Transcription factors bind sequence specifically in the promoter region upstream from the RNA polymerase start site and coding region to express a gene of interest. Polyamides have been shown to modulate the expression of downstream genes regulated by

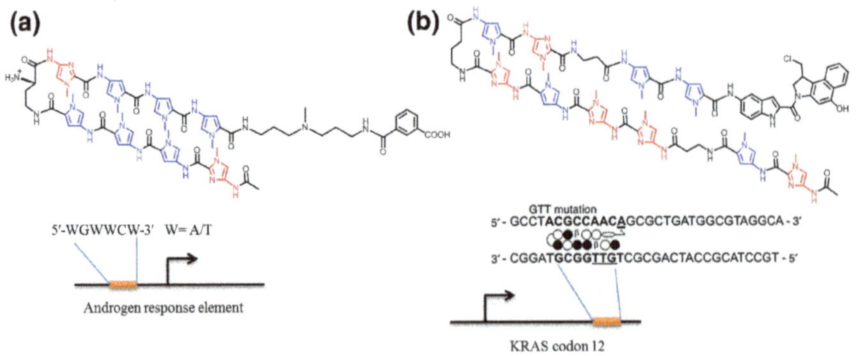

Fig. 1.6 **a** Polyamide targeting the androgen response element (ARE) found in the prostate-specific antigen (PSA) promoter inhibits expression of PSA as well as the transcripts. This polyamide has shown antitumor activity in prostate tumor xenograft model with limited host toxicity. **b** Polyamide conjugated with alkylating agents *Seco*-CBI targeting the KRAS codon 12 mutation. This polyamide has shown significant tumor growth suppression, with low host toxicity in KRAS-mutated but not wild-type tumors

glucocorticoid [25], androgen [26], and estrogen [27] by targeting at their respective consensus response elements. An example of antitumor activity by a hairpin polyamide was demonstrated by targeting 5′-WGWWCW-3′ (W = A/T) found in androgen response element in a prostate tumor xenograft model with limited host toxicity (Fig. 1.6) [28]. Another approach for specific gene inhibition was realized by designing alkylating agents conjugated with a sequence-specific polyamide targeting at the coding region of a gene [29]. Alkylating polyamide also ensures specific gene silencing ability in some oncogene, which is mutated at a specific site. Alkylating polyamides are designed by conjugating *Seco*-CBI (1,2,9,9a-tetrahydrocyclopropa [1,2-c]benz[1,2-e]indol-4-one), derived from the natural product duocarmycin, which alkylates Adenine at N3 position in a sequence-specific manner. Kras codon 12 mutation was targeted by this class of alkylation polyamides (Fig. 1.6) [30]. Alkylated polyamide can block RNA polymerase in the transcription process at the coding region that results in the truncated mRNAs.

1.2.2 PIPs as Gene Activator

PIPs were also designed to activate the transcription of a gene. To do that, the polyamide was attached to a transcription activator domain taken from proteins. Transcriptional activators are proteins that bind to specific DNA sequences and recruits transcriptional machinery to a proximal promoter, thereby stimulating gene expression. Relied on this concept, a hairpin polyamide attached with a short peptide of 20 amino acids act as an activation domain, for instance PEFPGIEL QELQELQALLQQ (AH) (20 mer) [31]. This artificial transcriptional activator was

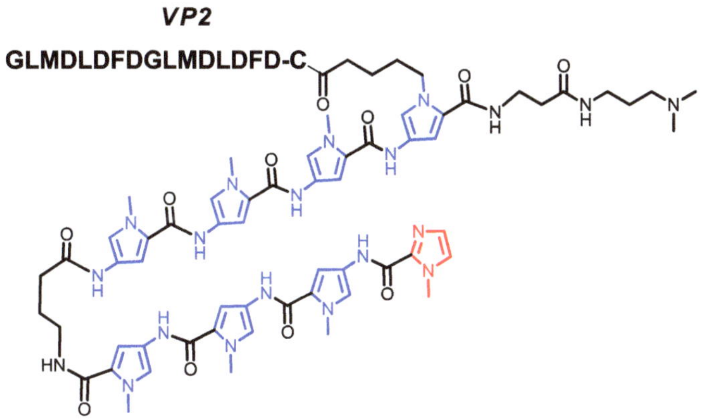

Fig. 1.7 The concept of transcriptional activation of gene by a polyamide–peptide conjugates. The hairpin polyamide DNA-binding domain binds to a specific promoter and the activation peptide domain VP16 binds and recruits parts of the transcription machinery

shown to stimulate promoter-specific transcription in a cell-free system. Later, size of the polyamide peptide conjugate was cut down from 20 mer peptide AH to 16 mer peptide DFDLDMLGDFDLDMLG which also results in the transcription activation. This peptide was derived from the activator domain of viral activator VP16 [32]. Transcriptional activation by peptide–polyamide conjugate is shown in Fig. 1.7.

Another transcriptional activator was reported which combined a hairpin polyamide, which can bind to the target DNA sequence and a wrench-shaped small molecule (wrenchnolol) that can bind to the Sur-2 protein, a subunit of human mediator complex that links transcription activators to RNA polymerase II (Fig. 1.8) [33]. This showed that the polyamide–wrenchnolol conjugate could activate the target gene but was limited due to poor cell permeability. Thus, it was possible to create synthetic transcription factor out of nonpeptidic component as well.

1.2.3 PIPs with HDAC Inhibitory Activity

Another approach for gene activation was considered by taking into account that epigenetic modification plays an important role in gene regulation. The eukaryotic genome is packaged into highly packaged chromatin architecture. Nucleosome is a fundamental unit of chromatin formed by histone octamer (H3, H4, H2A, and H2B) wrapping 146 base pairs of DNA [34]. In the tight chromatin structure, it is believed that genes are silent due to the inaccessibility of transcription factors. Chromatin modifications such as acetylation, methylation, phosphorylation, and ubiquitination are known to induce transcriptional activation. In some cases, methylation, sumoylation, deamination, and proline isomerization exhibit transcriptional repression [35]. In particular, histone acetylation is a dynamic process that regulates by two

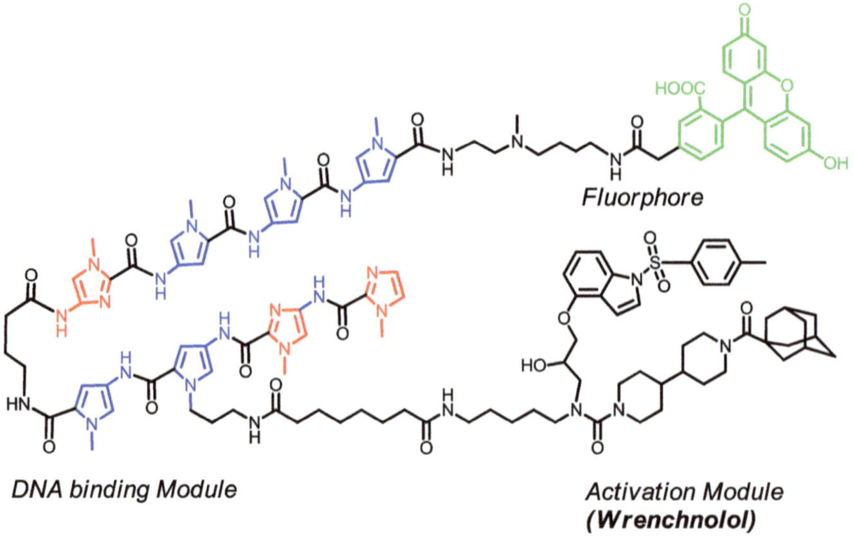

Fig. 1.8 The chemical structure of *wrenchnolol* conjugated polyamide as gene activator. This hybrid design takes the advantage of specific DNA-binding affinity of the polyamide and also the ability of *wrenchnolol* to bind sur-2 subunit of human mediator complex

large families of enzymes: (a) the histone acetylatransferase (HAT) and (b) the histone deacetylase transferase (HDAC). The balance between the action these two enzymes serve as a key regulatory mechanism for gene expression and govern several other developmental processes. The presence of acetylated lysine in histone tails is associated with relaxed chromatin state and gene-transcription activation; while the deacetylation of lysine residues increases the ionic interactions between the positively charged histones and negatively charged DNA yielding a more compact chromatin structure and represses gene transcription by limiting the accessibility of the transcription machinery. Owing to the fact that HDAC inhibitors have antitumor activity, hence it is considered as a potential therapeutic agent in recent days [36]. Trichostatin A and SAHA (suberoylanilide hydroxamic acid) are well known as epigenetic modifying histone deacetylase (HDAC) inhibitor and SAHA has received Food and Drug Administration approval for treating patients with cutaneous T-cell lymphoma. Although these HDAC inhibitors are promising, they act in sequence-independent manner resulting nonspecific gene activation (Fig. 1.9).

To realize specific gene activation, it is necessary to place a selective chromatin modifier that could precisely modulate the complex transcriptional network by exogenously introducing transcription factors. Hence, attachment of a HDAC inhibitor, SAHA, with a DNA-binding module, PI polyamide is an ideal solution to activate specific gene of interest. Interestingly, the recognition of polyamides was investigated in the core nucleosome structure resulting no binding hindrance for the polyamide [37]. Since polyamides can bind in the tight nucleosome structure, with the presence of SAHA it can open the tight structure and allow the transcription factor to activate the gene of interest.

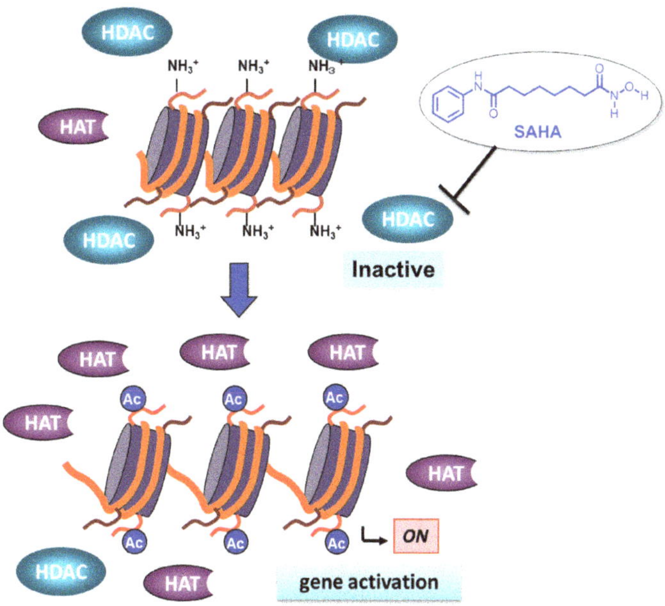

Fig. 1.9 HDAC inhibitors SAHA activate genes through acetylation of lysine residues of the histone tail. This results in loosening the tight chromatin structure aiding easy access for transcription factors

Therefore, for the first time a new compound, SAHA-PIP, was designed and synthesized for specific gene activation [38]. The first SAHA-PIP was designed to target the DNA sequence, 5′-WGCWGGC-3′ (W = A or T), present in the promoter region of the p16 tumor suppressor gene (Fig. 1.10). This new compound dramatically induced morphological changes in HeLa cells and selective Histone H3K9 acetylation.

Hence, inspired by this significant morphological changes suggesting selective gene-inducing ability of SAHA-PIP, a library of sixteen SAHA-PIPs from A to P as shown in Fig. 1.11 were synthesized [39]. Each individual SAHA-PIP from the library of 16 SAHA conjugates were treated in mouse embryonic fibroblast (MEF) cells and their effects on the expression of iPSC factors were screened [40]. Screening of the 16 SAHA-PIPs on MEF cells using pluripotency reprogramming factors as candidate genes revealed that compound E could activate mOct3/4 and mNanog by 3 fold, whereas mSox2, mc-Myc and mKlf4 activated around 1.5 fold [36]. However, the induction values were much lower than that of ES cell. Later on, various structural modifications were done on **E** to improve its biological activity [41]. But, structural modifications of **E** could not lead to any improvement in the gene expression. To this point the only choice was to redesign polyamide sequences by improving their sequence recognition ability, which could lead to the hit compound. Hence, a second generation of another sixteen SAHA-PIPs from Q to φ were synthesized (Fig. 1.11) [42]. In a similar set of experiment, each individual

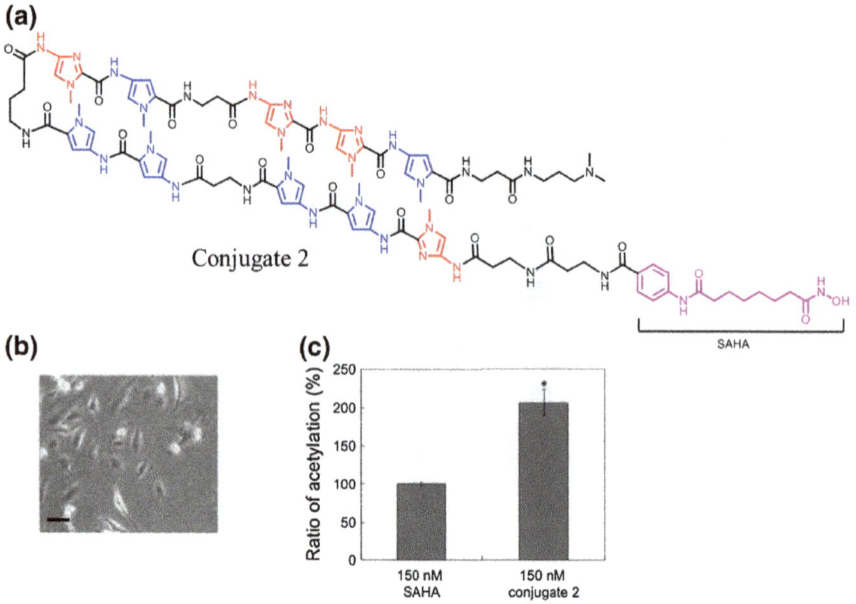

Fig. 1.10 **a** Chemical structure of the first SAHA-PIP targeting p16 tumor suppressor gene. SAHA unit shown in pink color conjugated at the N-terminus of the polyamide. **b** HeLa cells were cultured in complete Dulbecco's Modified Eagle's Medium (SIGMA, St. Louis, MO, USA) supplemented with 10% fetal bovine serum in 5% CO_2 at 37 °C with the following reagents: 150 nM SAHA conjugate 2 (**b**) for 2 weeks. Scale bar indicates 100 μm. **c** HeLa cells were treated with the indicated reagents (150 nM) for 7 days. After immunoprecipitation by acetylated (lysine 9) or whole H3 antibody, the amount of p16 promoter sequence in the co-precipitated DNAs was determined by quantitative PCR. The ratio of the amount of PCR products in acetylated/ whole immunoprecipitants in 150 nM SAHA-treated cells as 100% in this graph [38]

SAHA-PIP was screened from the second library on MEF using iPSC as candidate genes. Interestingly two distinct SAHA-PIPs, **Q** and **δ** activated some of the core pluripotency genes (*mOct3/4, mSox2, mNanog, mCdh1, mDppa4*) in MEF in just 100 nM (polyamide concentration) for 24 h treatment.

These 32 SAHA-PIPs were simultaneously screened on human dermal fibroblast HDF using the same iPSC candidate genes. In HDF-treated cell, a new SAHA-PIP known as **I** could activate pluripotency genes in just 1 μM for 48 h treatment [43]. In contrast, the compounds **E**, **Q**, **δ** which activated the same iPSC factor in MEF cell could not activate the same genes in HDF-treated cell. This result demonstrates the unique sequence selectivity of SAHA-PIP in gene regulation. To further support the above statement, there was one experimental proof by microarray analysis that the genome-wide gene expression of each individual SAHA-PIP triggers transcriptional activation of an exclusive cluster of genes and non-coding RNAs (Fig. 1.11) [44, 45]. The result was consistent in both microarray and qRT-PCR. Thus, each SAHA-PIP can target separate gene network based on its specific DNA

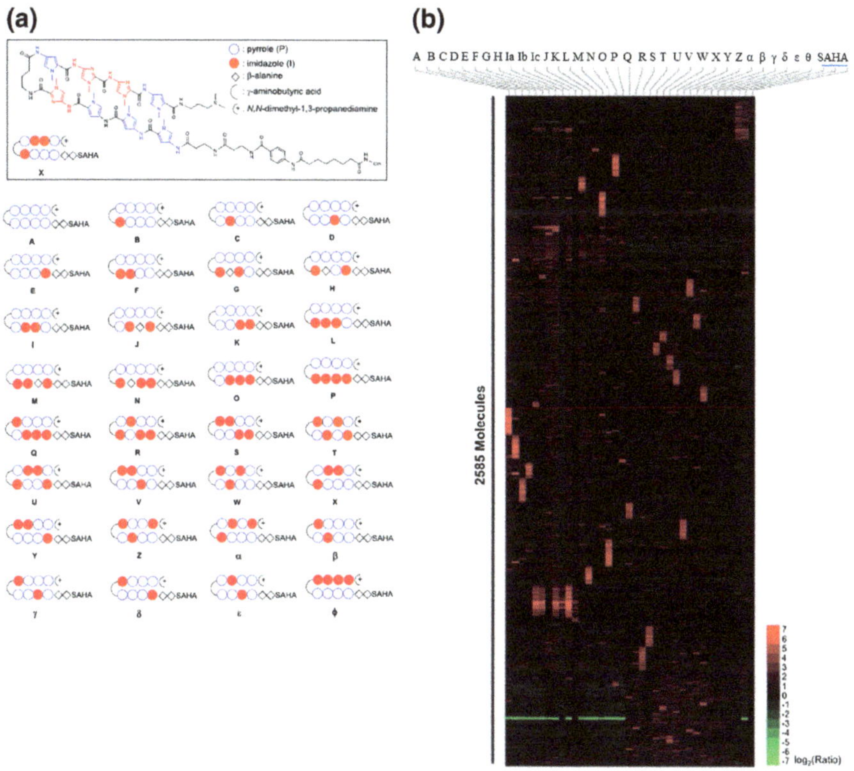

Fig. 1.11 **a** Design of 32 SAHA-PIPs. First library A–P and Second library Q–φ. **b** An unsupervised hierarchical clustering analysis of top 100 upregulated genes in SAHA-PIP 1–32 treated fibroblasts suggests that each SAHA-PIP activates a unique cluster of genes. Each result represents the sum of two individual culture plates. For SAHA-PIP 9 and SAHA, data derived from additional biological replicates is shown [44]

sequence recognition ability. Some examples of SAHA-PIPs that are characterized by further experiments on candidate genes of the respective gene networks are shown in Fig. 1.12.

1.3 Conclusion and Prospect

Generation of induced pluripotent stem cell from the somatic cell by the defined factors such as Oct3/4, Sox2, Klf4, and c-Myc pay much attention to many researchers due to its potential clinical applications as regenerative medicine. The possible risk of carcinogenesis such as potential of retroviruses to cause tumor in the tissues derived from host iPSC factors was thought to be a factor that could hamper its clinical translation. Thus, small molecule-based approach for the

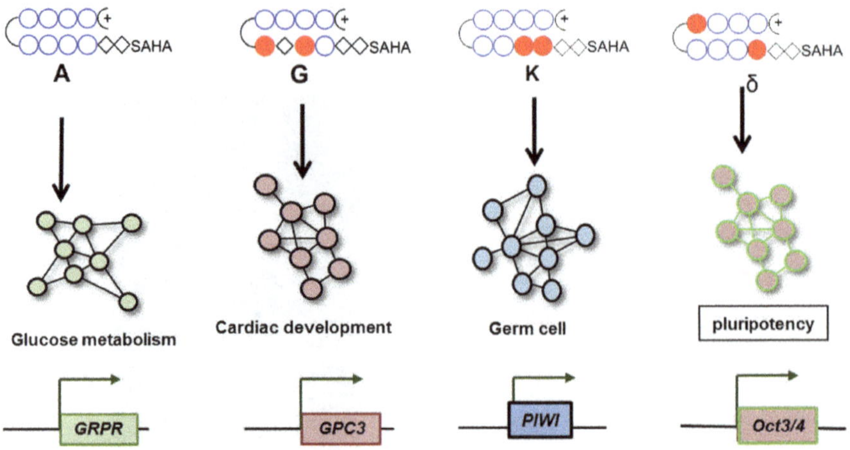

Fig. 1.12 Shown activation of different gene network by individual SAHA-PIPs such as A-activated GRPR pathway genes (glucose metabolism), G-activated cardiac developmental gene network, K-activated PIWI pathway gene (germ cell), and δ-activated pluripotency gene network (*Oct3/4* as the target gene)

transition of somatic cell to iPS cell is attractive due to lesser risk of carcinogenesis in clinical application. In this thesis from Chaps. 2 and 3, we have demonstrated the synthesis and biological activity of a SAHA-PIP, which activates core pluripotency genes and developmental genes in mouse embryonic fibroblast. We also demonstrated the synthesis protocol for improving the biological efficacy of this SAHA-PIP. Hence, for developing regenerative medicine, this small molecule could be a useful candidate.

1.4 General Introduction: (B) The Photochemistry of ^{Br}U-Substituted DNA

Modification of a native DNA sequence by replacing thymidine with 5-bromouracil (^{Br}U) greatly increases the photosensitivity of the resulting DNA. Creation of such photoreactive nucleic acid chromophore by replacing methyl group of thymidine with Br is attractive because the van der Waal radius of Br (1.95 Å) is similar to that of methyl group (2.00 Å). However, in the case of iodine, it is 2.15 Å, only 8% larger than the methyl group. Hence, this modification does not change the functionality of the resulting DNA, however it increases the photosensitivity with respect to protein–nucleic acid crosslinks [46], single- and double-strand breaks [47], and the creation of alkali-labile sites [48]. The DNA-mediated excess electron transfer (EET) is considered as the main photochemical events in the degradation or damage of ^{Br}U-substituted DNA. In the following section, the theories of DNA-mediated charge transfer are described based on two models namely oxidative hole transfer

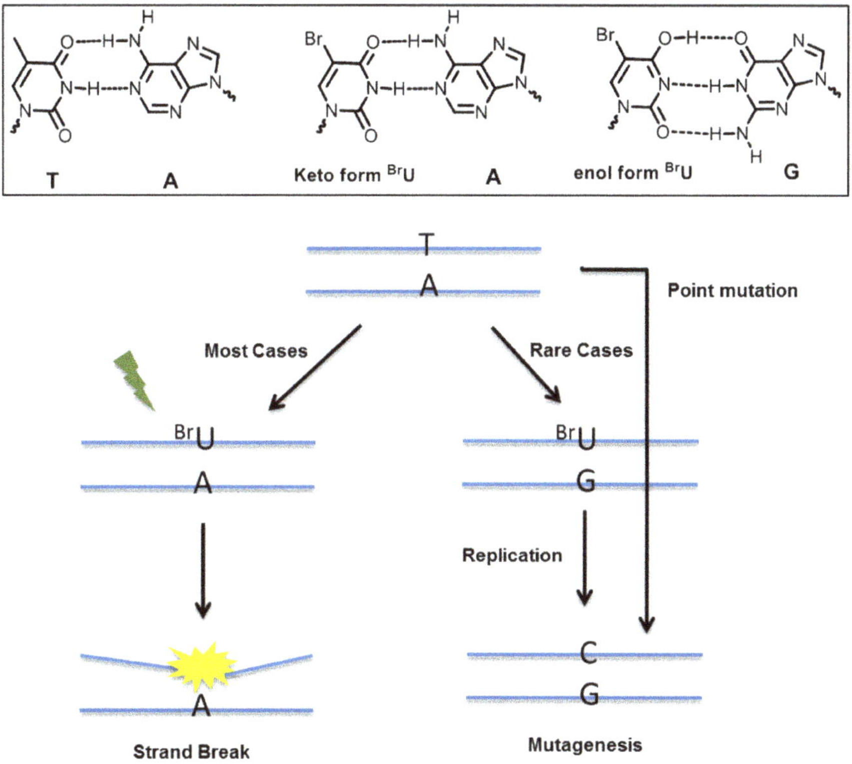

Fig. 1.13 Shown the isosteric structure of thymine and 5-bromouracil (^{Br}U). The keto form of ^{Br}U pairs with A similar to T with A. Thus, T can be easily replaced with ^{Br}U which is sensitive to UV. In rare cases the enol form of ^{Br}U pairs with G. This causes point mutation during the replication of DNA (the transition from A—T to G—C)

and reductive electron transfer. These studies were particularly important to understand DNA damage mechanism and also for the development of DNA nanotechnology devices. ^{Br}U substitution in DNA was used in clinical studies for cancer therapy due to its higher sensitivity to ionization radiation. In rare cases, ^{Br}U can also induce point mutation such as the transition from A—T to G—C (as shown in Fig. 1.13) [49]. The enol form of ^{Br}U can pair with G and during a subsequent round of replication this causes a mutation. This point mutation is toxic to cells. Over the years, the photochemistry of ^{Br}U-substituted DNA gained interest for the application in understanding various biological events under in vitro condition.

1.4.1 Hole Transfer Versus *Electron Transfer*

In principle, DNA-mediated charge transfer can be categorized as oxidative hole transfer and reductive electron transfer reactions. Both of these processes are in fact

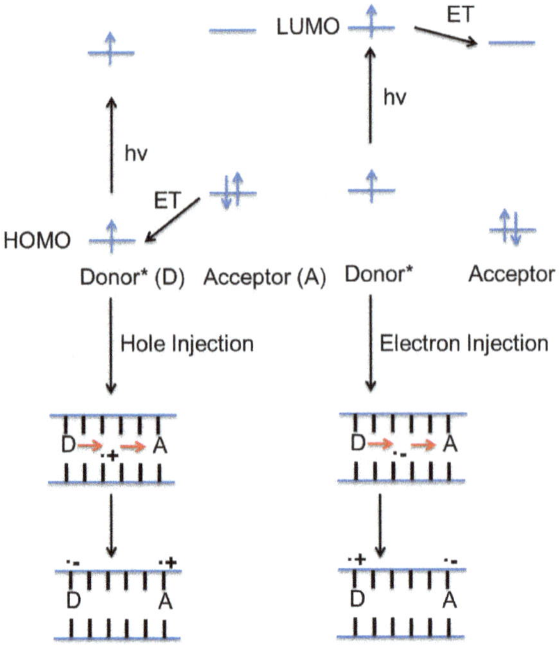

Fig. 1.14 Showing the photoinduced hole injection and oxidative hole transfer in HOMO-controlled *versus* electron injection and reductive electron transfer in a LUMO-controlled fashion

electron transfer reactions; however, the classification was done on the basis of their molecular orbital involved. During the oxidative hole transfer, an electron is transferred from the DNA or the final acceptor to the excited state of the donor in a HOMO-controlled fashion. In the case of reductive electron transfer, the electron of the excited state donor is transferred to the final acceptor in a LUMO-controlled fashion (Fig. 1.14) [50–52].

1.4.2 Mechanism of Oxidative Hole Transfer and Reductive Electron Transfer

The biological significance related to DNA damage was concerned mainly on oxidative hole transfer processes through DNA and hence many theories have been put forwarded [53]. A simplified model for oxidative hole transfer and reductive electron transfer is given in Fig. 1.15. Basically, two model systems have been proposed such as (1) super-exchange model and (2) hopping model for charge transfer in both categories such as oxidative hole transfer [54] and reductive electron transfer [50]. In a DNA-based donor–bridge–acceptor system, the energy level of the bridge to donor and acceptor (δE_D and δE_A) determines the charge

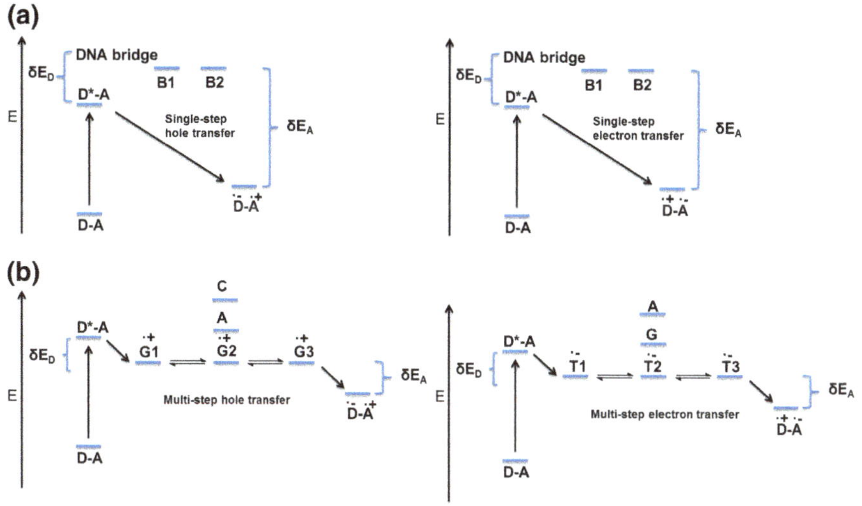

Fig. 1.15 Oxidative hole transfer and reductive electron transfer through DNA bridges via **a** superexchanged mechanism **b** by hopping mechanism. D = Donor, A = Acceptor, B = DNA base

transfer processes. For instance, if the bridge state of the medium DNA is energetically higher than the excited state of the donor, the charge will not be localized on the bridge. In contrast, if the energy level of the bridge is comparable to the excited state of the donor, charge transfer could occur through hopping process. This charge hopping process on the bridge is an intermediate step that occurs via several multi-step processes to the final charge trapping. Most frequently the bridge in hole transfer process is G, because it is easily oxidized and hence G^{-} serves as a hole carrier during hopping process [55]. The G hopping model was extended by A hopping model with longer A-T stretches, if Gs are absent in the sequential context [56]. In contrast, in case of reductive electron transfer, cytosine (C), thymine (T) or uracil are reduced more easily than Adenine (A) and G [57]. Hence, it was believed that pyrimidine radical anions of C^{-} and T^{-} play the role of electron carrier during electron-hopping process. However, the subsequent analysis suggests that C^{-} is not a good electron carrier.

1.5 Experimental Evidence of Charge Transfer

1.5.1 Hole Transfer

The hole transfer phenomena in DNA are of fundamental importance for DNA damage and DNA repair. By experimental evidence, it is shown that a positive charge can move through DNA over significant distances and this may define the damaged genome sites during oxidative stress. One of the first examples for

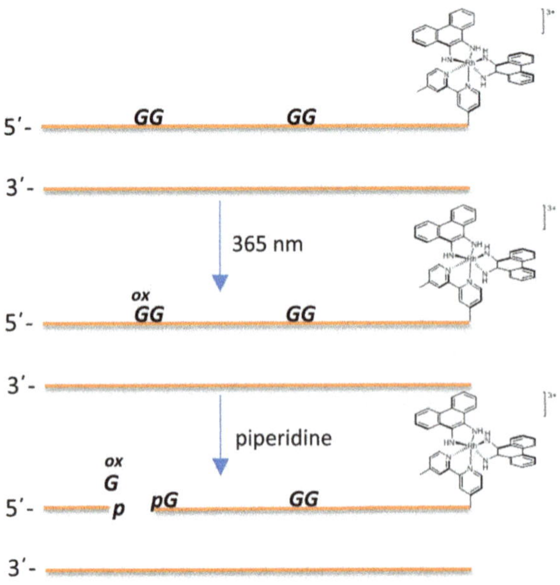

Fig. 1.16 Example of hole transfer from a distant in a duplex DNA. A rhodium intercalator was attached covalently to the one end of the DNA act as an electron acceptor and two 5′-GG-3′ doublets were included as an electron donor under photoexcitation at 365 nm. The site of guanine oxidation (G^{ox}) was cleaved by piperidine treatment [58]

oxidative hole transfer was shown using a metallo-intercalators to introduce a photoexcited hole into the DNA π-stack at a specific site in order to produce oxidative damage. A rhodium intercalator is covalently attached to one end of the DNA and separated spatially from 5′-GG-3′, the site of oxidation (Fig. 1.16) [58]. Since the guanine base is known to have the lowest ionization potential (IP) among the four DNA nucleobases, the stacking interaction of two consecutive guanine bases would create a site having an extremely low IP in duplex DNA. It has been shown that stacking of two guanine bases significantly lowers the IP and that the HOMO of the stacked 5′-GG-3′ is localized mainly on the 5′-G in B-form DNA [59].

1.5.2 Reductive Electron Transfer

Reductive ET processes are currently used in DNA chip technology [60] and DNA nanotechnology [61]. Reductive ET processes were studied based on two different kinetic electron traps, which are a specially designed T–T dimer or ^{Br}U. Some of the examples are consisting of an artificial DNA base with a flavine structure as the photoexcitable electron donor and a special T–T dimer as the electron trap [62]. It has been demonstrated that the amount of T–T dimer cleavage depends rather weakly on the distance to the flavine group, indicating an electron-hopping process. In contrast, some other examples demonstrated the reductive ET using a diaminonaphthalene derivative as the photoexcitable charge donor or phenothiazine (Pz)-modified 2′-deoxyuridine (Pz-dU) contains g DNA and ^{Br}U as the electron trap (Fig. 1.17) [63, 64]. In these studies, one similar observation was observed that the

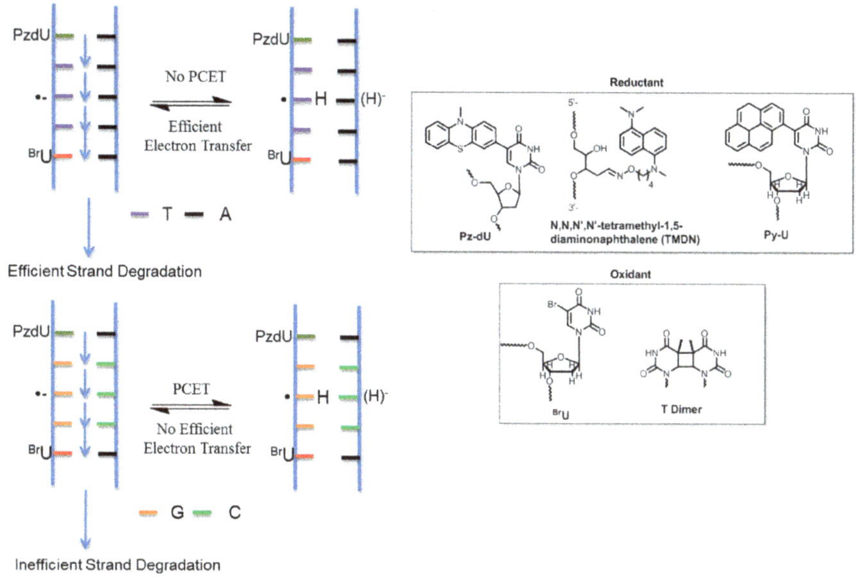

Fig. 1.17 A donor–acceptor-based reductive electron transfer system based on Pz-dU as the electron donor and ^{Br}U as the electron trap. In right some other examples of electron donor/reductant and an electron acceptor/oxidant that have been used before. (PCET: proton-coupled electron transfer) [52]

ET is dependent on the intervening base pairs. Because T–A as intervening base pairs serves better electron transfer efficiency than C–G. This is due to proton transfer which interferes with the ET in the later case, thus cytosine radical anion does not serve as a good electron carrier in a mixed DNA sequence.

1.6 Electron Transfer in ^{Br}U-Substituted DNA

1.6.1 Direct Irradiation and Intramolecular Excess Electron Transfer

Based on the above-mentioned reductive ET examples, which were found to be dependent on intervening base pairs between donor and acceptors, direct irradiation of a ^{Br}U-DNA also can be categorized as an example of reductive ET process. ^{Br}U-DNA when exposed to UV light (302 nm) gives sequence-specific strand cleavages. It was found that 2′-deoxyuridin-5-yl radicals (or uracil-5-yl radical) are effectively generated in $5'\text{-G/C[A]}^{\text{Br}}_{n=1,2,3}\text{U}^{\text{Br}}\text{U-3}'$, sequences in double-stranded DNA, resulting in the selective formation of heat labile 2′-deoxyribonolactone (Fig. 1.18). These sequences are called as hotspot sequences in the ^{Br}U-substituted DNA [65]. This intramolecular electron transfer is based on donor–acceptor model,

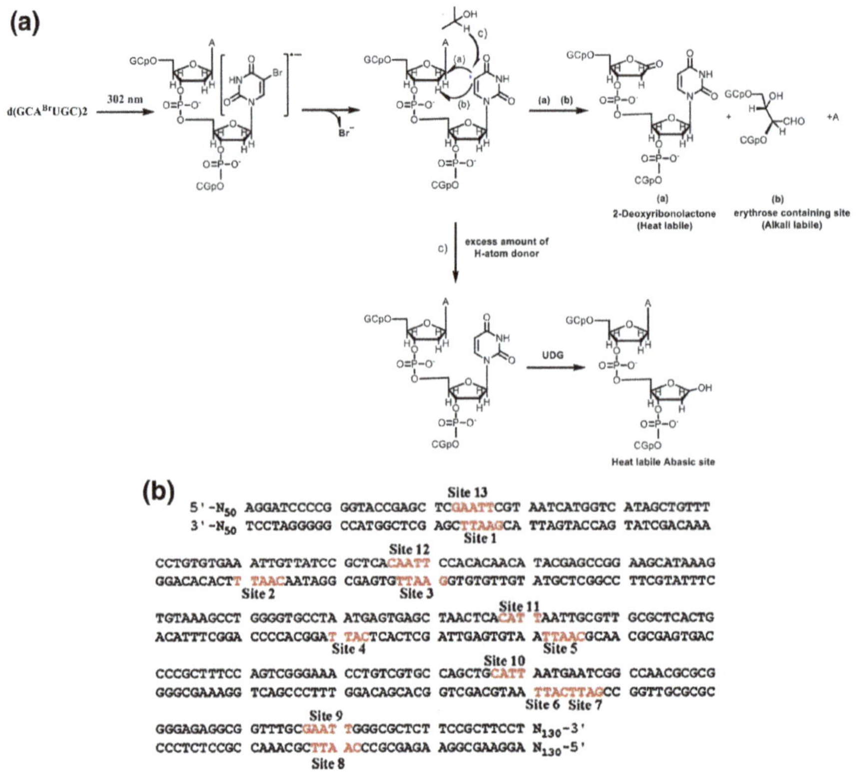

Fig. 1.18 **a** The scheme of strand cleavage is shown. Uracil-5-yl radical generates from uracil-5-yl anion after eliminating the bromide (Br^-). The putative electron donor is G from the opposite strand. In B-DNA, uracil-5-yl radical abstracts H-atom either from C1′-to generate 2-deoxyribonolactone (path a) or from C2′-α position to generate erythrose-containing sites (path b). The uracil-5-yl radical also can be quenched to U by the supplement of isopropanol which subsequent treatment of uracil DNA glycosylase (UDG) can be converted to abasic sites (path c). **b** The sequence of 450 bp DNA fragments was photo-irradiated using 302 nm. The cleavage sites are shown in red which are known as hotspot sequences. All Ts were substituted with either ^{Br}U or ^{I}U [65b]

where G acts as a donor and ^{Br}U as acceptor and consecutive A–Ts acts as a bridge. In these hotspot sequences, A–T bridges play an important role for charge carrier. It is to be noted that without such bridges this electron transfer does not occur at all in 5′-G^{Br}U^{Br}U-3′ sequences. The mechanism of strand cleavage was established after several subsequent studies and it is found that generation of uracil-5-yl radical is the main photochemical event. This radical is a powerful H-atom abstractor and subsequently abstracts a hydrogen atom from the C1′ position of the nearest sugar part at the 5′-end to form heat labile 2-deoxyribonolactone. Another simultaneous hydrogen abstraction by uracil-5-yl radical was also detected from C2′-αH of the same sugar moiety to form alkali-labile erythrose-containing sites [66]. This hydrogen abstraction by the radical species can be quenched by the supplement of excess hydrogen atom donor such as THF or isopropanol in the reaction mixture [65d].

More precisely, this hydrogen abstraction by uracil-5-yl radical was found to be atom specific and highly dependent on the local DNA conformations; such as A-form, B-form, Z-form, bent DNA, and G-quadruplexes [67]. For example, competitive C1′- and C2′α-hydrogen abstractions have been observed in B-DNA, whereas selective C1′-hydrogen abstraction occurs in the A-like structure of DNA–RNA hybrids. In Z-form DNA, stereospecific C2′β-hydrogen abstraction gives rise to C2′α-hydroxylation. In protein-induced DNA kinks, photoirradiation causes intrastrand hydrogen abstraction from the 5-methyl group of thymine on the 5′side. The photoreactivity of iodouracil-containing telomeric DNA depends on the orientation of the G-quadruplex. The 2′-deoxyribonolactone residue is effectively produced only in the diagonal loop of the antiparallel G-quadruplex.

1.6.2 Ligand-Sensitized Excess Electron Transfer to ^{Br}U-Substituted DNA

In order to use the photochemistry of ^{Br}U-DNA for developing DNA-based detection assay, ligand-sensitized electron transfer to ^{Br}U was proposed. To achieve this, an organic chromophore which can be easily photooxidized was attached to the ligand. The DNA minor groove binder, polyamides, was attached with pyrene as an electron donor. This intermolecular system of donor and acceptor gives an excellent yield of photoproduct under UVA irradiation condition (Fig. 1.19). This proof of concept of intermolecular electron transfer on a short oligo opens up the possibility of developing a photo-footprinting technique for detecting the binding sites of DNA-binding small molecule in a ^{Br}U-substituted DNA [68].

1.6.3 Excess Electron Transfer from Protein to ^{Br}U-Substituted DNA

A short peptide *Lys–Trp–Lys* was used to repair the thymine dimer upon UV exposure [69]. The photoinduced excess electron transfer from the tryptophan residue of the peptide does the cycloreversion of T dimer. Some reports also suggest that photolyase repairs thymine dimer by electron transfer from cofactor domain under the action of UV light [70]. It is also found that photolyase can repair thymine dimers without the involvement of a cofactor. This suggests that specific amino acid itself has the potential to repair DNA by a photoinduced electron transfer mechanism [71]. Some reports suggest that the guanine-cation radical formed upon oxidative hole transfer is reduced by electron transfer from histone [72]. A direct evidence of excess electron transfer was from Sso7d, a nucleohistone-like protein of archaea, to ^{Br}U or thymine dimer in which a tryptophan was involved as the electron donor (Fig. 1.20) [73]. Considering this specific

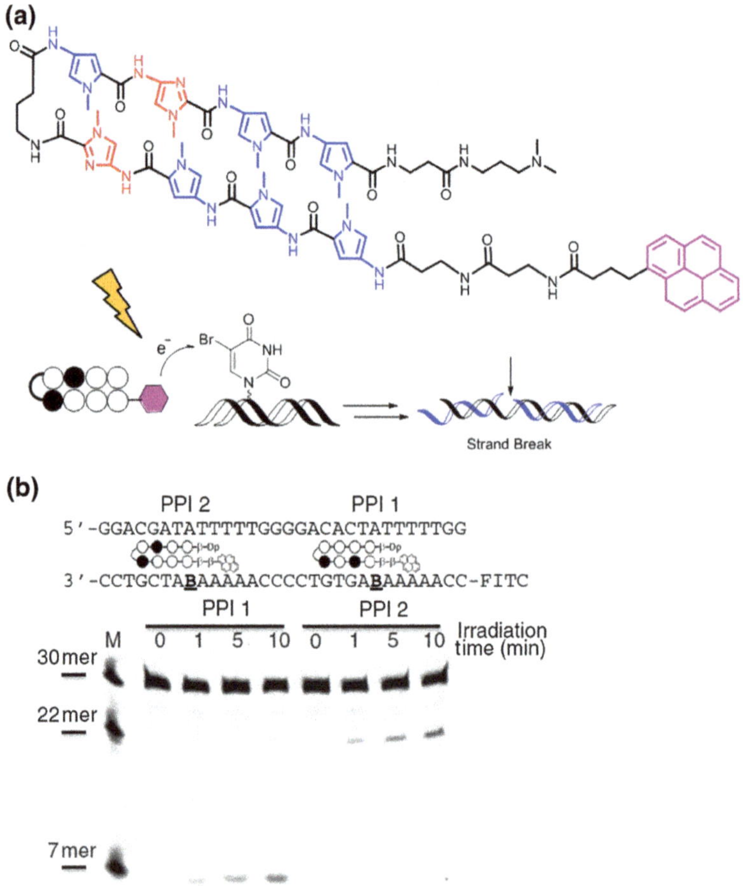

Fig. 1.19 **a** Chemical structure of pyrene-conjugated pyrrole–imidazole polyamide and sequence-specific electron injection. **b** Site-specific electron injection into DNA. An oligonucleotide containing two ^{Br}U residues and FITC was prepared and photo-irradiated with PPI 1 or PPI 2 for the indicated time on ice. The resulting samples were cleaved at the uracil site using UNG following heat treatment and loaded onto gels. A 7-mer, 22-mer, and 30-mer ODNs were used as a size marker [68]

amino acid as the source of electron in proteins, it can be further exploited to investigate the complex biological events such as multi-protein complex formation during transcription process on ^{Br}U-substituted DNA.

1.7 Conclusion and Future Prospect

Based on the strand cleavage chemistry, an overview of the various applications of ^{Br}U-substituted DNA in the past and present can be summarized in the following Fig. 1.21. In 1990, first by direct irradiation of ^{Br}U-DNA the damage was studied

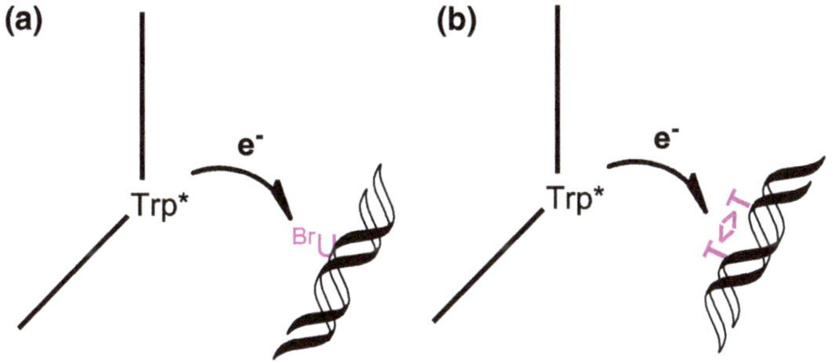

Fig. 1.20 Electron transfer from the excited state of Trp-24 of Sso7d to **a** DNA containing ^{Br}U residues and **b** the cycloreversion of thymine dimer upon 280 nm irradiation

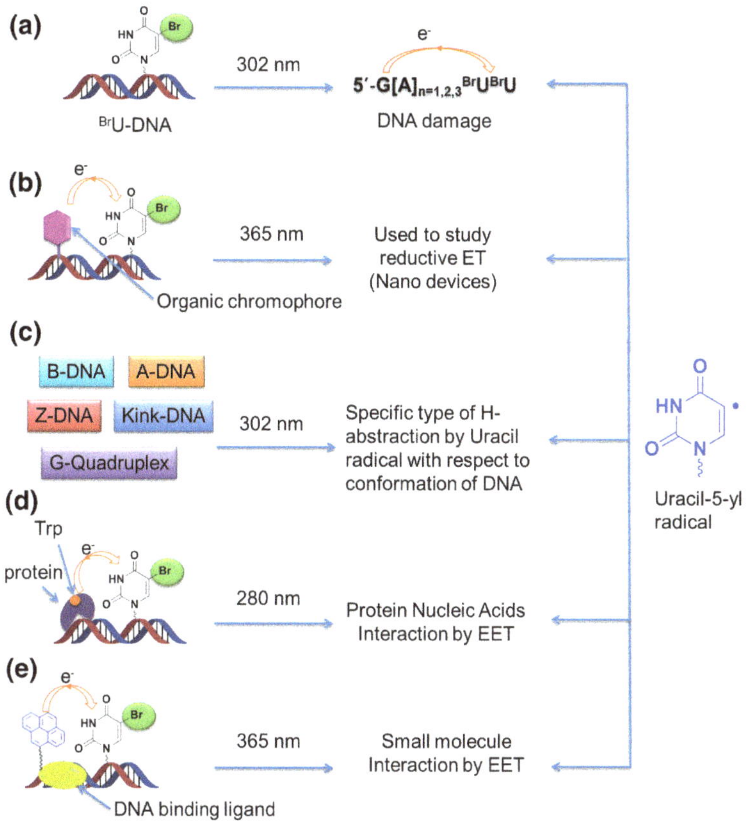

Fig. 1.21 Overview of various applications of ^{Br}U-labeled DNA. **a** The study of strand cleavage and identification of hotspot sequences on ^{Br}U-DNA. **b** Study of reductive ET. **c** The study of strand cleavage on A-DNA, B-DNA, Z-DNA, G-quadruplex, and DNA-kink by EET. **d** The study protein-sensitized strand cleavage on ^{Br}U-DNA by EET and **e** the study of intermolecular electron transfer using PIPs

based on photodegraded product of six mer oligo d(GCA^{Br}UGC). This was the first step toward the understanding of DNA damage in ^{Br}U-DNA. From 1990 to 2005, from six mer to 450 bp, the journey has contributed significant developments to this field. Specific sequences, called hotspots, were identified on the long DNA, which are highly sensitive to damage. The damage occurred via an interesting intramolecular electron transfer process as described in the Sect. 1.6.1. ^{Br}U-DNA was also used to study the mechanism of photoinduced charge transfer by connecting an organic chromophore (with absorbance at around 350 nm) on the DNA. These studies were devoted to understand how distance and sequence dependence charge migration could occur and this would also help in developing DNA-based nanodevices. Later on, DNA damage was studied on the different DNA structures such as A-DNA, Z-DNA, DNA-kink, and G-quadruplex and compared with B-DNA. This gives another interesting aspect of photochemical reaction which suggests that the H-abstraction is dependent on local conformations of the DNA resulting in specific photoproducts. Simultaneously, protein-sensitized strand cleavage on ^{Br}U-DNA was investigated to identify protein–nucleic acids interactions by excess electron transfer. This tool offers a new concept to investigate complex biological interactions of proteins with nucleic acids under in vitro condition. Owing to the fact that ^{Br}U-DNA remains undamaged under 365 nm irradiation, DNA-binding small molecule with organic chromophore was used for intermolecular electron transfer.

Combing the previous contributions in this field, it is possible to develop ^{Br}U-DNA-based assay to screen small molecules and proteins by EET. In Chap. 4, we have demonstrated a photo-footprinting method to screen the binding affinity of pyrene-conjugated pyrrole–imidazole polyamides by EET. In Chap. 5, we have demonstrated the cooperative binding affinity of two-transcription factors Sox2 and Pax6 on a ^{Br}U-labeled regulated element DC5 by EET. In Chap. 6, the Hoechst 33,258 induced double strand on ^{Br}U-labeled DNA is demonstrated. These successful studies open up the further application of it in studying the small molecule and protein interactions in secondary structures of DNA.

References

1. Travers AA (1989) DNA conformation and protein binding. Annu Rev Biochem 58:427–452. https://doi.org/10.1146/annurev.bi.58.070189.002235; (b) Wang AH, Quigley GJ, Kolpak FJ, Crawford JL, van Boom JH, van der Marel G, Rich A (1979) Molecular structure of a left-handed double helical DNA fragment at atomic resolution. Nature 282:680–686; (c) Sinden RR (1994) DNA structure and function. Academic Press, New York; (d) Sinden RR, Pearson CE, Potaman VN, Ussery DW (1998) Advances in Genome Biology, 5:1–141. https://doi.org/10.1016/S1067-5701(98)80019-3
2. (a) Rich A, Zhang S (2003) Timeline: Z-DNA: the long road to biological function. Nat Rev Genet 4:566–572. https://doi.org/10.1038/nrg1115; (b) Hackett JA, Feldser DM, Greider CW (2001) Telomere dysfunction increases mutation rate and genomic instability. Cell 106: 275–286

3. Dervan PB (2001) Molecular recognition of DNA by small molecules. Bioorg Med Chem 9:2215–2235. https://doi.org/10.1016/S0968-0896(01)00262-0
4. Weaver RF. Molecular biology, 5th edn. (2011) DNA-protein interactions in bacteria, pp 222–238 (Chapter 9)
5. Brennan RG, Matthews BW (1989) The helix-turn-helix DNA binding motif. J Biol Chem 264:1903–1906
6. Pavletich NP, Pabo CO (1991) Zinc finger-DNA recognition: crystal structure of a Zif268-DNA complex at 2.1 Å. Science 252:809–817
7. Pomerantz JL, Sharp PA (1994) Homeodomain determinants of major groove recognition. Biochemistry 33:10851–10858
8. König P, Richmond TJ (1993) The X-ray structure of the GCN4-bZIP bound to ATF/CREB site DNA shows the complex depends on DNA flexibility. J Mol Biol 233:139–154. https://doi.org/10.1006/jmbi.1993.1490
9. Baraldi PG, Bovero A, Fruttarolo F, Preti D, Tabrizi MA, Pavani MG, Romagnoli R (2004) DNA minor groove binders as potential antitumor and antimicrobial agents. Med Res Rev 4:475–528
10. (a) Van Dyke MW, Hertzberg RP, Dervan PB (1982) Map of distamycin, netropsin, and actinomycin binding sites on heterogeneous DNA: DNA cleavage-inhibition patterns with methidiumpropyl-EDTA.Fe(II). Proc Natl Acad Sci USA 79:5470; (b) Van Dyke MW, Dervan PB (1983) Chromomycin, mithramycin, and olivomycin binding sites on heterogeneous deoxyribonucleic acid. Footprinting with (methidiumpropyl-EDTA)iron(II). Biochemistry 22:2373; (c) Van Dyke MW, Dervan PB (1984) Echinomycin binding sites on DNA. Science 225:1122; (d) Schultz PG, Taylor JS, Dervan PB (1982) Design synthesis of a sequence-specific DNA cleaving molecule. (Distamycin-EDTA)iron(II). J Am Chem Soc 104:6861; (d) Taylor JS, Schultz PG, Dervan PB (1984) DNA affinity cleaving. sequence specific cleavage of DNA by Distamycin-EDTA·Fe(II) and EDTA-Distamycin·Fe(II) Tetrahedron 40:457–465; (e) Dervan PB (1986) Design of sequence-specific DNA-binding molecules. Science 232:464; (f) Meier JL, Yu AS, Korf I, Segal DJ, Dervan PB (2012) Guiding the design of synthetic DNA-binding molecules with massively parallel sequencing. J Am Chem Soc 134:17814–17822; (g) Anandhakumar C, Kizaki S, Bando T, Pandian GN, Sugiyama H (2015) Advancing small-molecule-based chemical biology with next-generation sequencing technologies. Chem BioChem 16:20–38. https://doi.org/10.1002/cbic.201402556
11. (a) Lerman LS (1961) Structural considerations in the interaction of DNA and acridines. J Mol Biol 3:18; (b) Müller W, Crothers DM (1968) Studies of the binding of actinomycin and related compounds to DNA. J Mol Biol 35:251–290; (c) Bresloff JL, Crothers DM (1975) DNA-ethidium reaction kinetics: demonstration of direct ligand transfer between DNA binding sites. J Mol Biol 95:103; (d) Waring MJ, Wakelin FPC (1974) Echinomycin: a bifunctional intercalating antibiotic. Nature 252:653; (e) Arcamone F, Penco S, Orezzi P, Nicolella V, Pirelli A (1964) Structure and synthesis of distamycin A. Nature 203:1064
12. Kopka ML, Yoon C, Goodsell D, Pjura P, Dickerson RE (1985) The molecular origin of DNA-drug specificity in netropsin and distamycin. Proc Natl Acad Sci USA 82:1376
13. Pelton JG, Wemmer DE (1989) Structural characterization of a 2:1 distamycin A.d (CGCAAATTGGC) complex by two-dimensional NMR. Proc Natl Acad Sci USA 86: 5723–5727
14. Lown JW, Krowicki K, Bhat UG, Skorobogaty A, Ward B, Dabrowiak JC (1986) Molecular recognition between oligopeptides and nucleic acids: novel imidazole-containing oligopeptides related to netropsin that exhibit altered DNA sequence specificity. Biochemistry 25:7408
15. Wade WS (1989) Ph.D. Thesis, California Institute of Technology
16. Wade WS, Mrksich M, Dervan PB (1992) Design of peptides that bind in the minor groove of DNA at 5′-(A, T)G(A, T)C(A, T)-3′ sequences by a dimeric side-by-side motif. J Am Chem Soc 114:8783. https://doi.org/10.1021/ja00049a006
17. Mrksich M, Parks ME, Dervan PB (1994) Hairpin peptide motif. A new class of oligopeptides for sequence-specific recognition in the minor groove of double-helical DNA. J Am Chem Soc 116:7983. https://doi.org/10.1021/ja00097a004

18. De Cian A, Delemos E, Mergny JL, Teulade-Fichou MP, Monchaud D (2007) Highly efficient G-quadruplex recognition by bisquinolinium compounds. J Am Chem Soc 129: 1856–1857. https://doi.org/10.1021/ja067352b
19. Rodriguez R, Müller S, Yeoman JA, Trentesaux C, Riou JF, Balasubramanian S (2008) A novel small molecule that alters shelterin integrity and triggers a DNA-damage response at telomeres. J Am Chem Soc 130:15758–15759. https://doi.org/10.1021/ja805615w
20. Chung WJ, Heddi B, Hamon F, Teulade-Fichou MP, Phan AT (2014) Solution structure of a G-quadruplex bound to the bisquinolinium compound Phen-DC(3). Angew Chem Int Ed Engl 53:999–1002. https://doi.org/10.1002/anie.201308063
21. Herman DM, Baird EE, Dervan PB (1998) Stereochemical control of the DNA binding affinity, sequence specificity, and orientation preference of chiral hairpin polyamides in the minor groove. J Am Chem Soc 120:1382. https://doi.org/10.1021/ja9737228
22. Herman DM, Turner JM, Baird EE, Dervan PB (1999) Cycle polyamide motif for recognition of the minor groove of DNA. J Am Chem Soc 1999(121):1121
23. Turner JM, Swalley SE, Baird EE, Dervan PB (1998) Aliphatic/aromatic amino acid pairings for polyamide recognition in the minor groove of DNA. J Am Chem Soc 120:6219. https://doi.org/10.1021/ja980147e
24. Herman DM, Baird EE, Dervan PB (1999) Tandem hairpin motif for recognition in the minor groove of DNA by pyrrole-imidazole polyamides. Chem Eur J 5:975–983
25. Muzikar KA, Nickols NG, Dervan PB (2009) Repression of DNA-binding dependent glucocorticoid receptor-mediated gene expression. Proc Natl Acad Sci USA 106:16598–16603. https://doi.org/10.1073/pnas.0909192106
26. Nickols NG, Dervan PB (2007) Suppression of androgen receptor-mediated gene expression by a sequence-specific DNA-binding polyamide. Proc Natl Acad Sci USA 104:10418–10423. https://doi.org/10.1073/pnas.0704217104
27. Nickols NG, Szablowski JO, Hargrove AE, Li BC, Raskatov JA, Dervan PB (2013) Activity of a Py-Im polyamide targeted to the estrogen response element. Mol Cancer Ther 12:675–684. https://doi.org/10.1158/1535-7163
28. Yang F, Nickols NG, Li BC, Marinov GK, Said JW, Dervan PB (2013) Antitumor activity of a pyrrole-imidazole polyamide. Proc Natl Acad Sci USA 110:1863–1868. https://doi.org/10.1073/pnas.1222035110
29. Bando T, Sugiyama H (2006) Synthesis and biological properties of sequence-specific DNA-alkylating pyrrole-imidazole polyamides. Acc Chem Res 39:935–944. https://doi.org/10.1021/ar030287f
30. Hiraoka K, Inoue T, Taylor RD, Watanabe T, Koshikawa N, Yoda H, Shinohara K, Takatori A, Sugimoto H, Maru Y, Denda T, Fujiwara K, Balmain A, Ozaki T, Bando T, Sugiyama H, Nagase H (2015) Inhibition of KRAS codon 12 mutants using a novel DNA-alkylating pyrrole-imidazole polyamide conjugate. Nat Commun 6:6706. https://doi.org/10.1038/ncomms7706
31. Mapp AK, Ansari AZ, Ptashne M, Dervan PB (2000) Activation of gene expression by small molecule transcription factors. Proc Natl Acad Sci USA 97:3930
32. Ansari AZ, Mapp AK, Nguyen DH, Dervan PB, Ptashne M (2001) Towards a minimal motif for artificial transcriptional activators. Chem Biol 8:583
33. Kwon Y, Arndt HD, Mao Q, Choi Y, Kawazoe Y, Dervan PB, Uesugi M (2004) Small molecule transcription factor mimic. J Am Chem Soc 126:15940–15941. https://doi.org/10.1021/ja0445140
34. Kornberg RD, Lorch Y (1999) Twenty-five years of the nucleosome, fundamental particle of the eukaryote chromosome. Cell 98:285–294
35. Pandian GN, Sugiyama H (2012) Programmable genetic switches to control transcriptional machinery of pluripotency. Biotechnol J 7:798–809. https://doi.org/10.1002/biot.201100361

36. Pontiki E, Lintina DH (2012) Histone deacetylase inhibitors (HDACIs). Structure-activity relationships: history and new QSAR perspectives. Med Res Rev 32:1–165. https://doi.org/10.1002/med.20200
37. Gottesfeld JM, Melander C, Suto RK, Raviol H, Luger K, Dervan PB (2001) Sequence-specific recognition of DNA in the nucleosome by pyrrole-imidazole polyamides. J Mol Biol 309:615–629. https://doi.org/10.1006/jmbi.2001.4694
38. Ohtsuki A, Kimura MT, Minoshima M, Suzuki T, Ikeda M, Bando T, Nagase H, Shinohara K, Sugiyama H (2009) Synthesis and properties of PI polyamide—SAHA conjugate. Tetrahedron Lett 50:7288–7292. https://doi.org/10.1016/j.tetlet.2009.10.034
39. Pandian GN et al (2011) Synthetic small molecules for epigenetic activation of pluripotency genes in mouse embryonic fibroblasts. Chem Biochem 12:2822–2828. https://doi.org/10.1002/cbic.201100597
40. Takahashi K, Yamanaka S (2006) Induction of pluripotent stem cells from mouse embryonic and adult fibroblast cultures by defined factors. Cell 126:663–676. https://doi.org/10.1016/j.cell.2006.07.024
41. Pandian GN, Ohtsuki A, Bando T, Sato S, Hashiya K, Sugiyama H (2012) Development of programmable small DNA-binding molecules with epigenetic activity for induction of core pluripotency genes. Bioorg Med Chem 20:2656–2660. https://doi.org/10.1016/j.bmc.2012.02.032
42. Pandian GN, Nakano Y, Sato S, Morinaga H, Bando T, Nagase H, Sugiyama H (2012) A synthetic small molecule for rapid induction of multiple pluripotency genes in mouse embryonic fibroblasts. Sci Rep 2:544. https://doi.org/10.1038/srep00544
43. Pandian GN, Sato S, Anandhakumar C, Taniguchi J, Takashima K, Syed J, Han L, Saha A, Bando T, Nagase H, Sugiyama H (2014) Identification of a small molecule that turns ON the pluripotency gene circuitry in human fibroblasts. ACS Chem Biol 9:2729–2736. https://doi.org/10.1021/cb500724t
44. Pandian GN, Taniguchi J, Junetha S, Sato S, Han L, Saha A, AnandhaKumar C, Bando T, Nagase H, Vaijayanthi T, Taylor RD, Sugiyama H (2014) Distinct DNA-based epigenetic switches trigger transcriptional activation of silent genes in human dermal fibroblasts. Sci Rep 4:3843. https://doi.org/10.1038/srep03843
45. Han L, Pandian GN, Junetha S, Sato S, Anandhakumar C, Taniguchi J, Saha A, Bando T, Nagase H, Sugiyama H (2013) A synthetic small molecule for targeted transcriptional activation of germ cell genes in a human somatic cell. Angew Chem Int Ed 52:13410–13413. https://doi.org/10.1002/anie.201306766
46. (a) Willis MC, Hicke BJ, Uhlenbeck OC, Cech TR, Koch TH (1993) Photocrosslinking of 5-iodouracil-substituted RNA and DNA to proteins. Science 262:1255–1257; (b) Hicke BJ, Willis MC, Koch TH, Cech TR (1994) Telomeric protein-DNA point contacts identified by photo-cross-linking using 5-bromodeoxyuridine. Biochemistry 33:3364–3373; (c) Ogata R, Gilbert W (1977) Contacts between the lac repressor and the thymines in the lac operator. Proc Natl Acad Sci USA 74:4973
47. Suzuki K, Yamauchi M, Oka Y, Suzuki M, Yamashita S (2011) Creating localized DNA double-strand breaks with microirradiation. Nat Protoc 6:134–139. https://doi.org/10.1038/nprot.2010.183
48. (a) Krasin F, Hutchinson F (1978) Strand breaks and alkali-labile bonds induced by ultraviolet light in DNA with 5-bromouracil in vivo. Biophys J 24:657–664; (b) Krasin F, Hutchinson F (1978) Double-strand breaks from single photochemical events in DNA containing 5-bromouracil. Biophys J. 24:645–656. https://doi.org/10.1016/s0006-3495(78)85410-1; (c) Sugiyama H, Tsutsumi Y, Fujimoto K, Saito I (1993) Photoinduced deoxyribose C2′ oxidation in DNA. Alkali-dependent cleavage of erythrose-containing sites via a retroaldol reaction. J Am Chem Soc 115:4443–4448. https://doi.org/10.1021/ja00064a004

49. Davidson RL, Broeker P, Ashman CR (1988) DNA base sequence changes and sequence specificity of bromodeoxyuridine-induced mutations in mammalian cells. Proc Natl Acad Sci USA 85:4406–4410
50. Giese B (2002) Long-distance electron transfer through DNA. Annu Rev Biochem 71:51–70. https://doi.org/10.1146/annurev.biochem.71.083101.134037
51. Wagenknecht HA (2003) Reductive electron transfer and transport of excess electrons in DNA. Angew Chem Int Ed 42:2454–2460. https://doi.org/10.1002/anie.200301629
52. Prunkl C, Berndl S, Wanninger-Weiss C, Barbaric J, Wagenknecht HA (2010) Photoinduced short-range electron transfer in DNA with fluorescent DNA bases: lessons from ethidium and thiazole orange as charge donors. Phys Chem Chem Phys 12:32–43. https://doi.org/10.1039/b914487k
53. Schuster G (ed) (2004) Longe-range charge transfer in DNA I. Springer, Berlin
54. Jortner J, Bixon M, Langenbacher T, Michel-Beyerle ME (1998) Charge transfer and transport in DNA. Proc Natl Acad Sci USA 95:12759–12765
55. Steenken S, Jovanovic SV (1997) How easily oxidizable is DNA? one-electron reduction potentials of adenosine and guanosine radicals in aqueous solution. J Am Chem Soc 119:617–618. https://doi.org/10.1021/ja962255b
56. Giese B, Amaudrut J, Köhler AK, Spormann M, Wessely S (2001) Direct observation of hole transfer through DNA by hopping between adenine bases and by tunnelling. Nature 412:318–320. https://doi.org/10.1038/35085542
57. (a) Seidel CAM, Schulz A, Sauer MHM (1996) Nucleobase-specific quenching of fluorescent dyes. 1. Nucleobase one-electron redox potentials and their correlation with static and dynamic quenching efficiencies. J Phys Chem 100:5541–5553. https://doi.org/10.1021/jp951507c; (b) Steenken S, Telo JP, Novais HM, Candeias LP (1992) One-electron-reduction potentials of pyrimidine bases, nucleosides, and nucleotides in aqueous solution. Consequences for DNA redox chemistry. J Am Chem Soc 114:4701–4709. https://doi.org/10.1021/ja00038a037
58. Hall DB, Holmlin RE, Barton JK (1996) Oxidative DNA damage through long-range electron transfer. Nature 382:731–735. https://doi.org/10.1038/382731a0
59. Sugiyama H, Saito I (1996) Theoretical studies of GG-specific photocleavage of DNA via electron transfer: significant lowering of ionization potential and 5′-localization of HOMO of stacked GG bases in B-Form DNA. J Am Chem Soc 118:7063–7068. https://doi.org/10.1021/ja9609821
60. Boon EM, Salas JE, Barton JK (2002) An electrical probe of protein–DNA interactions on DNA-modified surfaces. Nat Biotechnol 20:282–286. https://doi.org/10.1038/nbt0302-282
61. Niemeyer CM, Adler M (2002) Nanomechanical devices based on DNA. Angew Chem Int Ed Engl 41:3779–3783. https://doi.org/10.1002/1521-3773(20021018)41:20<3779:AID-ANIE3779>3.0.CO;2-F
62. Haas C, Kräling K, Cichon M, Rahe N, Carell T (2004) Excess electron transfer driven DNA repair does not depend on the transfer direction. Angew Chem Int Ed Engl 43:1842–1844. https://doi.org/10.1002/anie.200353067
63. Ito T, Rokita SE (2003) Excess electron transfer from an internally conjugated aromatic amine to 5-bromo-2-deoxyuridine in DNA. J Am Chem Soc 125:11480–11481. https://doi.org/10.1021/ja035952u
64. Wagner C, Wagenknecht HA (2005) Reductive electron transfer in phenothiazine-modified DNA is dependent on the base sequence. Chem-Eur J 11:1871–1876. https://doi.org/10.1002/chem.200401013
65. (a) Sugiyama H, Tsutsumi Y Saito I (1990) Highly sequence-selective photoreaction of 5-bromouracil-containing deoxyhexanucleotides. J Am Chem Soc 112:6720–6721. https://doi.org/10.1021/ja00174a046; (b) Watanabe T, Bando T, Xu Y, Tashiro R Sugiyama H (2005) Efficient generation of 2′-deoxyuridin-5-yl at 5′-(G/C)AA(X)U(X)U-3′ (X = Br, I)

sequences in duplex DNA under UV irradiation. J Am Chem Soc 127:44–45. https://doi.org/10.1021/ja0454743; (c) Watanabe T, Tashiro R Sugiyama H (2007) Photoreaction at 5′-(G/C) AABrUT-3′ Sequence in duplex DNA:efficient generation of uracil-5-yl radical by charge transfer. J Am Chem Soc 129:8163–8168. https://doi.org/10.1021/ja0692736; (d) Hashiya F, Saha A, Kizaki S, Li Y, Sugiyama H (2014) Locating the uracil-5-yl radical formed upon photoirradiation of 5-bromouracil-substituted DNA. Nucleic Acids Res 42:13469–13473. https://doi.org/10.1093/nar/gku1133

66. Sugiyama H, Fujimoto K, Saito I (1996) Evidence for intrastrand C2′ hydrogen abstraction in photoirradiation of 5-halouracil-containing oligonucleotides by using stereospecifically C2′-deuterated deoxyadenosine. Tetrahedron Lett 37:1805–1808. https://doi.org/10.1016/0040-4039(96)00123-2
67. Xu Y, Tashiro R, Sugiyama H (2007) Photochemical determination of different DNA structures. Nat Protoc 2:78–87. https://doi.org/10.1038/nprot.2006.467
68. Morinaga H, Takenaka T, Hashiya F, Kizaki S, Hashiya K, Bando T, Sugiyama H (2013) Sequence-specific electron injection into DNA from an intermolecular electron donor. Nucleic Acids Res 41:4724–4728. https://doi.org/10.1093/nar/gkt123
69. BehmoarasT Toulme JJ, Hélène C (1981) A tryptophan-containing peptide recognizes and cleaves DNA at apurinic sites. Nature 292:858–859. https://doi.org/10.1038/292858a0
70. (a) DeRosa MC, Sancar A, Barton JK (2005) Electrically monitoring DNA repair by photolyase. Proc Natl Acad Sci USA 102:10788–10792. https://doi.org/10.1073/pnas.0503527102; (b) Boon EM, Livingston AL, Chimiel NH, David SS, Barton JK (2003) DNA-mediated charge transport for DNA repair. Proc Natl Acad Sci USA 100:12543–12547. https://doi.org/10.1073/pnas.2035257100; (c) Yavin E, Boal AK, Stemp ED, Boon EM, Livingston AL, O'Shea VL, David SS, Barton JK (2005) Protein-DNA charge transport: redox activation of a DNA repair protein by guanine radical. Proc Natl Acad Sci USA 102: 3546–3551. https://doi.org/10.1073/pnas.0409410102
71. Kim S, Li Y, Sancar A (1992) The third chromophore of DNA photolyase: Trp-277 of Escherichia coli DNA photolyase repairs thymine dimers by direct electron transfer. Proc Natl Acad Sci USA 89:900–904
72. Cullis PM, Jones GDD, Symons MCR, Lea JS (1987) Electron transfer from protein to DNA in irradiated chromatin. Nature 330:773–774. https://doi.org/10.1038/330773a03340144
73. Tashiro R, Wang AH, Sugiyama H (2006) Photoreactivation of DNA by an archaeal nucleoprotein Sso7d. Proc Natl Acad Sci USA 103:16655–16659. https://doi.org/10.1073/pnas.0603484103

Chapter 2
Synthesis and Biological Evaluation of Targeted Transcriptional Activator with HDAC8 Inhibitory Activity

Abstract We demonstrated the activation of genes that played important roles in the early development process of living animals. To do that, we developed a new class of compounds known as SAHA-PIP, which is a combination of two functional domains. One is DNA minor groove-binding pyrrole–imidazole polyamide and the other one is histone deacetylase inhibitor HDACi called SAHA. In a previous study, we developed Sδ as one of such compound exhibiting both DNA-binding and HDAC-inhibitory activity. Epigenetic activity of Sδ was attributed to the active metal-binding (–NHOH) domain of SAHA. We synthesized a derivative of Sδ, called Jδ to evaluate the role of surface recognition domain (–phenyl) of SAHA in Sδ-mediated transcriptional activation. In vitro studies revealed that Jδ displayed potent inhibitory activity against HDAC8. Jδ retained the pluripotency gene-inducing ability of Sδ when used alone and in combination with Sδ; a notable increase in the pluripotency gene expression was observed. Interestingly, Jδ significantly induced the expression of HDAC8-controlled Otx2 and Lhx1. Our results suggest that the epigenetic activity of our multifunctional molecule could be altered to improve its efficiency as a transcriptional activator for intricate gene network(s).

Keywords Biomimetic synthesis · DNA recognition · HDAC8 inhibition Transcriptional activators · Developmental genes

2.1 Introduction

DNA minor groove-binding *N*-methylpyrrole (Py)-*N*-methylimidazole (Im) polyamides (PIPs) are used to learn the mechanism of cancer [1, 2]. The reason is these molecules can recognize each of the four Watson–Crick base pair sequences through side-by-side stacked-ring pairing: Im/Py distinguishes G–C from C–G, while the Py/Py pairing is degenerate and targets A–T and T–A [3]. As they can recognize predetermined DNA sequences with high specificity and high binding ability, which is similar to the natural transcription factor, thus PIPs have

A. Saha, *Molecular Recognition of DNA Double Helix*, Springer Theses,
https://doi.org/10.1007/978-981-10-8746-2_2

been extensively studied for regulating gene expression [4–6]. In recent years, these PIPs have been conjugated with different functional moieties including alkylating agents such as seco-CBI [7–10], and fluorescent dyes [11–16], for different biological activities including gene regulation. In nature, gene regulation is achieved at various distinctive levels, and programmable gene-based transcriptional activators have been overlooking the critical epigenetic influence in gene control. Coordinated epigenetic modifications precisely orchestrate the genome-wide gene expression to cause transcriptional activation and control cell fate. Several small molecules such as BIX-01294 [17], valproic acid [18], 5′Azacytidine [19] have been shown to artificially induce epigenetic modifications, however, they act in a sequence-independent manner. SAHA (suberoylanilide hydroxamic acid) is one such epigenetic modifying histone deacetylase (HDAC) inhibitor that binds directly to the catalytic site of the class I and class II HDAC enzymes to induce transcriptionally permissive chromatin. SAHA is currently in advanced clinical trials for the treatment of cancer and has also been shown to be effective against latently HIV-infected T-cells [20].

To achieve better efficacy and biological properties, SAHA was conjugated with the sequence-specific PIPs termed as SAHA-PIP. Initially, a SAHA-PIP was designed to target the promoter region of the p16 tumor-suppressor gene, which induced site-specific histone H3 Lys9 acetylation and significant morphological changes in HeLa cells [20]. Encouraged by its selective inducing ability, we synthesized a library of 32 (A–ϕ) SAHA-PIPs with differential gene-inducing ability. HDAC inhibitors including SAHA are known to increase the reprogramming efficiency of somatic cells to induced pluripotent stem cells (iPSCs). Therefore, we chose iPSC factors as the candidate genes to evaluate the effect of SAHA-PIPs in mouse embryonic fibroblasts (MEF) [21]. We identified some SAHA-PIPs could distinctively activate the iPSC factors (Oct-3/4, Nanog, Sox2, Klf4, and c-Myc) by screening them on stem cell-related genes [22]. Among these 32 compounds, a SAHA-PIP termed as Sδ dramatically induced multiple pluripotency genes by more than tenfold to initiate cellular reprogramming in just 24 h [23, 24]. The mechanism regarding the selective induction of pluripotency genes by Sδ in mouse fibroblast suggests its sequence specificity among the others. Here, in this chapter, we focused on the modification of Sδ, especially in the SAHA functional domain to modulate HDAC-inhibitory activity. The goal is to have specific HDAC inhibition by modifying the original compound Sδ. This study is one step further to prove HDAC inhibition by SAHA-PIPs plays a significant role in its specific gene-inducing ability.

In that context, first we look at the X-ray crystallographic data based on the interaction of SAHA and homolog of HDAC [histone deacetylase-like protein (HDLP)], which revealed that SAHA inhibits HDACs to induce acetylated histones via three functional domains: an active metal-binding domain (–NHOH), a linker—$(CH2)_6$—domain, and a surface-recognition domain (-phenyl) [25, 26]. Among these domains, the metal-binding domain is crucial for gene expression. As without the active metal-binding domain in Sδ, this compound could not express the same genes [24]. However, the influence of other two domains, linker and surface-recognition domain, in Sδ is unknown. A recent report showed that subtle manipulations of SAHA like the addition of a ferrocene unit to generate a new class of compounds called JAHA (Jay Amin hydroxamic acid) could successfully alter its HDAC specificity [27]. Based on this study, we generated a new compound called Jδ (JAHA-PIP δ) by attaching the PIP δ with SAHA that does not contain the surface-recognition domain but retains the other two domains (Fig. 2.1). First, we demonstrate through studies with in vitro enzymatic assay for HDAC that the derivative Jδ possesses notably a higher HDAC8 inhibitory activity than that of the original Sδ. We also demonstrate a notable increase in the gene expression with the combination of Sδ and Jδ to suggest the importance of HDAC8 inhibition in pluripotency gene expression. HDAC8 are known to epigenetically control the skull morphogenesis by the repression of Otx2 and Lhx1 [28]. However, mode of expressing these development-related genes is challenging and no small molecules have been known to increase the expression of these critical genes. Herein, we report for the first time, the ability of our new type of synthetic small molecule to induce the endogenous expression of the HDAC8-regulated Otx2 and Lhx1 via hyperacetylation of histone H3 in their promoter region.

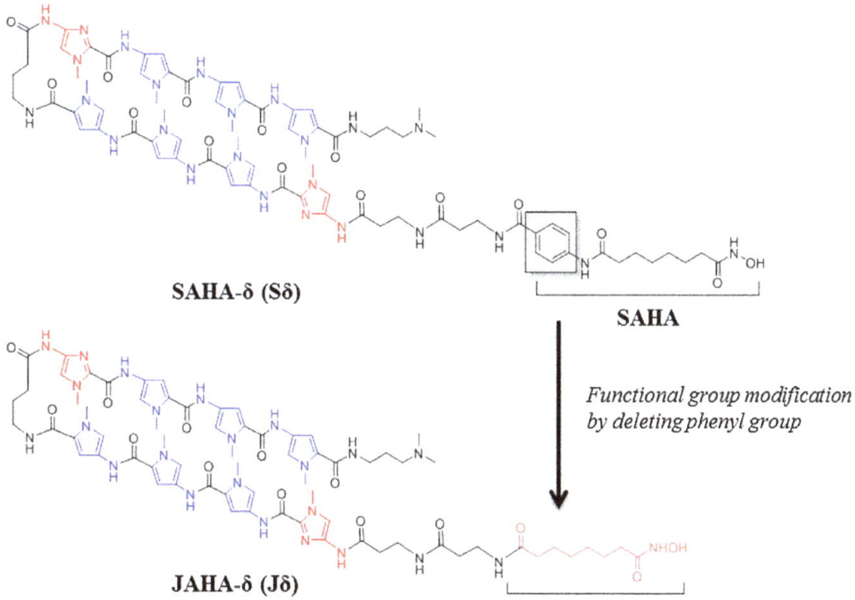

Fig. 2.1 Chemical structure of SAHA-δ and JAHA-δ

2.2 Results

2.2.1 Design of the Synthetic Transcriptional Activator

We designed the new compound by deleting the phenyl group from original SAHA-PIP (Sδ) [24]. Structure of our target compound Jδ is shown in Fig. 2.1. Jδ polyamides were synthesized by Fmoc solid-phase peptide synthesis using Py-oxime resin (Fig. 2.2). Using Fmoc chemistry, PIP-δ (NH_2-*β*-*β*-ImPyPyP-*γ*-ImPyPyPy-Dp) sequence was synthesized by PSSM-8 peptide synthesizer (Shimadzu, Kyoto) with a computer-assisted operating system with 45 mg of oxime resin (ca. 0.2 mmol/g, 200–400 mesh). After the solid-phase synthesis, the precursor was cleaved from the resin using *N*,*N*-dimethylaminopropylamine (Dp), stirred at 45 °C for 3 h. The reaction mixture was filtered, triturated from Et_2O, to yield as a yellow crude powder. The precursor A (about 24 mg) was dissolved in DMF (0.5 ml) and added 10 µl of DIEA (3 equiv.) and finally methyl-8-cholo-8-oxo-octanoate about 8 µl was added very slowly at room temperature and stirred for 1 h. This allowed the formation of precursor B which then converted to the active functional group of SAHA-PIP by using an aqueous solution of 50% (v/v) NH_2OH (0.5 ml). The reaction mixture was stirred for 8 h at room temperature. After the reaction, hydroxylamine was quenched with acetic acid (0.5 ml) at 0 °C. The mixture was purified by flash column chromatography. The purity was checked by HPLC (elution with trifluoroacetic acid and a 0–100% acetonitrile linear gradient (0–40 min) at a flow rate of 1.0 mL min^{-1} under 254 nm) to yield Jδ as a light yellow powder (2.0 mg, 14%). 1H NMR for the compound JAHA-δ ESI-TOF-MS (positive) m/z calculated for $C_{69}H_{90}N_{24}O_{14}$ $[M + 2H]^{2+}$ 740.3612; found 740.3617. 1H NMR (600 MHz, [D6]DMSO): δ = 10.34 (brs, 1H), 10.28 (s, 1H), 10.26 (s, 1H), 10.01 (s, 2H), 9.94 (s, 3H), 9.90 (s, 2H), 9.24 (brs, 1H), 8.15 (brt, 2H), 8.03 (brs, 2H), 7.96 (s, 1H), 7.77 (s, 1H), 7.45 (s, 1H), 7.43 (s, 1H), 7.26 (s, 1H), 7.22 (s, 1H), 7.20 (s, 1H), 7.16 (s, 1H), 7.14 (s, 1H), 7.12 (s, 1H), 7.06 (s, 1H), 7.03 (s, 1H), 6.94 (s, 1H), 6.90 (s, 1H), 3.95 (s, 1H), 3.85 (s, 1H), 3.84 (s, 1H), 3.80 (s, 1H), 3.79 (s, 1H), 3.42 (s, 1H), 3.41 (s, 1H), 3.38 (s, 1H), 2.78 (s, 1H), 2.38 (s, 1H),2.20 (s, 1H), 2.00 (t, 2H), 1.91 (t, 2H), 1.82 (m, 2H), 1.79 (m, 2H), 1.44 (m, 2H), 1.18 (m, 2H).

ESI-TOF-MS (positive) m/z calculated for $C_{69}H_{90}N_{24}O_{14}$ $[M+2H]^{2+}$ 740.3612; found 740.3617.

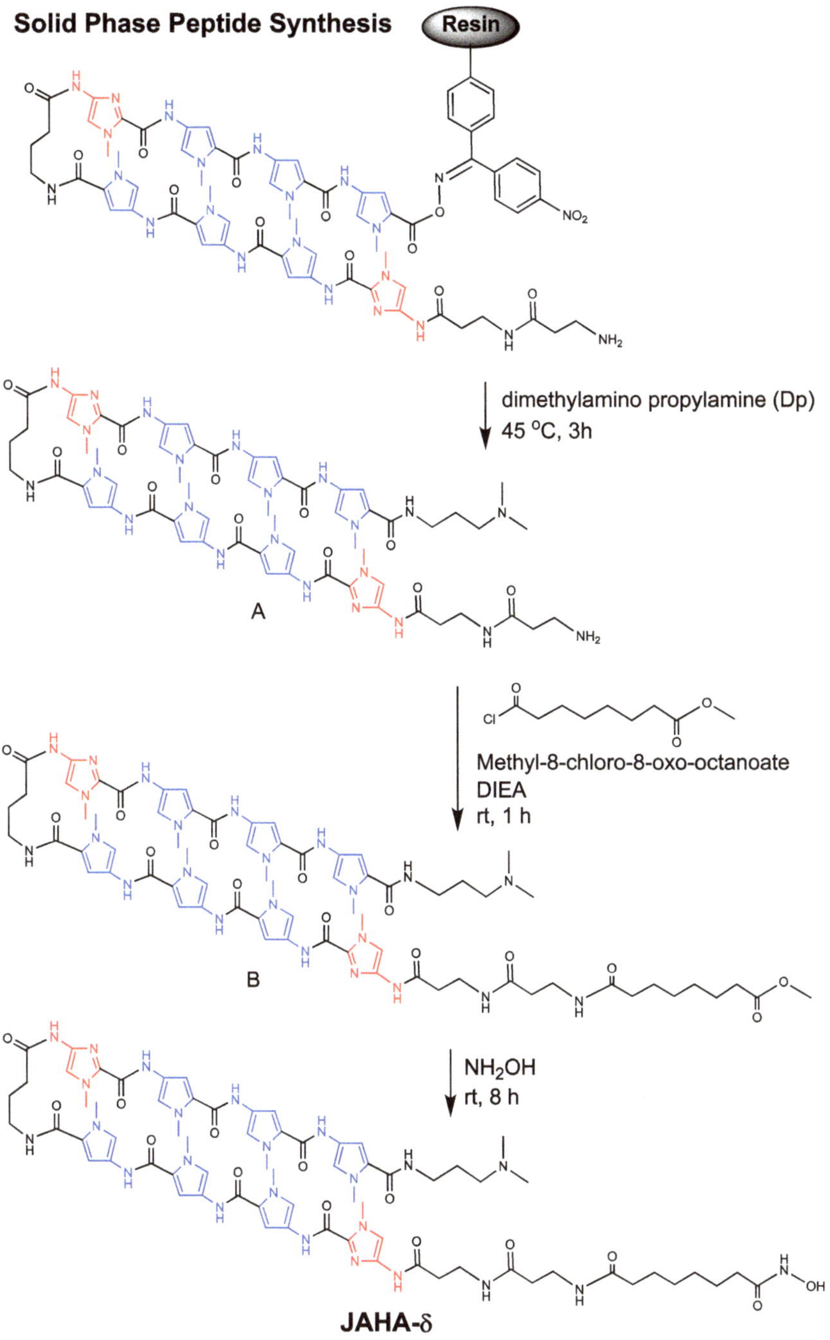

Fig. 2.2 Design and synthesis of a pyrrole–imidazole polyamide SAHA conjugate (Sδ) derivative termed as Jδ

2.2.2 *Determination of HDAC Activity* in Vitro

18 HDAC enzymes have been found in human and these are divided into three classes based on their homology to yeast deacetylases (Fig. 2.3) [25]. Class I includes HDAC1, 2, 3, and 8 with molecular weights of 22–55 kDa and homogenous in their catalytic site, Class II includes HDAC4, 5, 7, and 9 with molecular weight 120–135 kDa. A subclass of HDAC, class IIa includes HDAC6 and 10 containing two catalytic sites and HDAC11 is referred to as class IIb or class IV because it conserved residues in the catalytic core region shared by both class I and class II. Class III belongs to Sir2 family (SIRT1-7) that does not have histone as a primary target. Class I, II, and IV are zinc-dependent enzymes bearing a highly conserved catalytic domain with a Zn^{2+} ion [25].

It is identified that some HDAC enzymes play specific role in the development process. For example, HDAC6 promotes survival and regrowth of neurons following injury [29], HDAC1 regulates pluripotency [30], HDAC4 regulates skeletogenesis [31], and also HDAC8 controls skull morphogenesis [28]. SAHA is a pan-HDAC inhibitor that has a relatively higher specificity against class I HDACs [32]. Sδ was previously shown to epigenetically induce the endogenous expression of multiple pluripotency genes. However, the actual specificity of Sδ against individual HDAC enzymes was not evaluated. Here, we used a fluorogenic-based assay kit that contains individually purified HDAC enzymes such as HDAC1, HDAC2, or HDAC8, to evaluate HDAC activity of Sδ and Jδ in a 384-well plate using trichostatin (TSA) as the control [27]. IC_{50} values were normalized to a control sample by logistic regression and triplicate values were plotted (Fig. 2.4). In this assay, Jδ and Sδ exhibited similar potency against HDAC1 with IC_{50} values of 0.37 and 0.49 μM respectively. Against HDAC2, Sδ showed a relatively higher potency than that of Jδ (Table 2.1). Interestingly, a notable and substantial

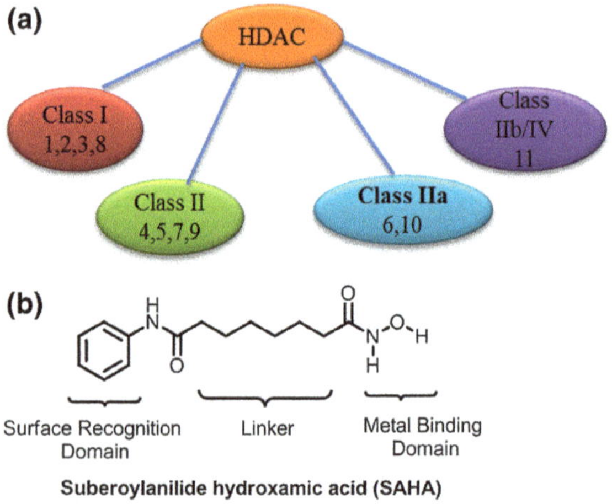

Fig. 2.3 **a** Different classes of HDAC enzyme are shown with their family partners and their specific role. **b** Structure of SAHA shown with different functional domains

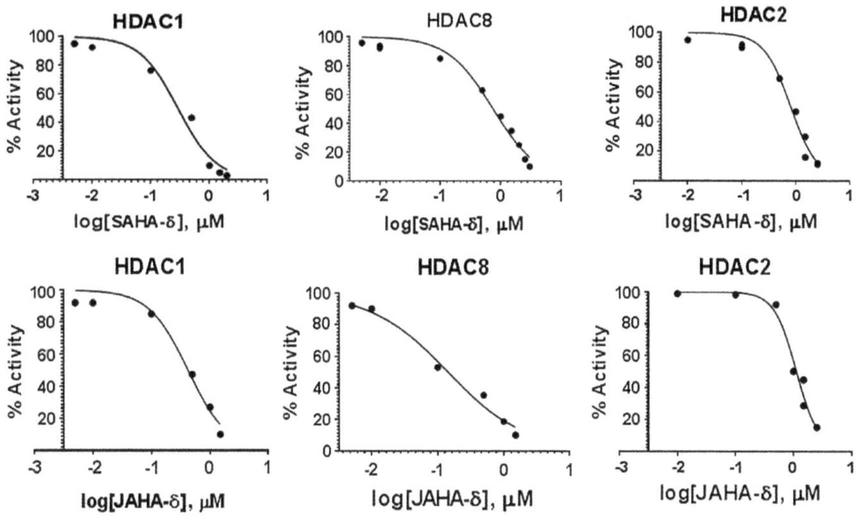

Fig. 2.4 HDAC activity for the compounds Sδ and Jδ

Table 2.1 IC_{50} values for HDAC inhibitors[b]

Test inhibitor	HDAC1[a] (μM)	HDAC2[a] (μM)	HDAC8[a] (μM)
Jδ	0.49 ± 0.07	1.12 ± 0.4	0.13 ± 0.06
Sδ	0.37 ± 0.09	0.77 ± 0.18	0.82 ± 0.1
Trichostatin A	0.0043 ± 0.001	0.014 ± 0.03	1.36 ± 0.2

[a]IC_{50} values for the compounds (μM) with SD in parenthesis
[b]In vitro enzymatic activity assay for HDAC1, 2, and 8 enzymes by trichostatin (TSA), Sδ, and Jδ with their IC_{50} values. Compounds were arrayed in a 384-well plate format as library stock solutions, with a maximum concentration of 5 mM and were serially diluted. HDAC enzymatic reactions were conducted at 37 °C for 30 min followed by the addition of a fluorogen. Enzymatic activity was determined by fluorometry at 350–380 nm excitation and 440–460 nm emission. The data comprise the mean of three individual experiments. Curves were fitted by logistic regression using Graph Pad Prism

difference in IC_{50} values was observed in HDAC8 inhibition by both the compounds. Jδ was found to be about 6 times more specific toward HDAC8 than that of Sδ. Jδ and Sδ displayed an IC_{50} value of about 0.13 and 0.82 μM respectively. It is important to note here that when compared with the TSA, Jδ was about ten times higher specific toward HDAC8 (Table 2.1 and Fig. 2.4). However, TSA showed a superior specificity against HDAC1 and HDAC2 than that of Jδ and Sδ. Taken together, our result supports the notion that the modifications in the surface-recognition domain could alter the specificity of SAHA moiety in SAHA-PIP against different HDACs (here HDAC8).

2.2.3 Gene Expression

In vitro HDAC activity analysis indicated Jδ to be more HDAC8 specific than Sδ. Also, its specificity against HDAC1 activity remains in the same range as Sδ. Because HDAC1 is known to regulate pluripotency [30], we performed screening studies in MEF to determine the effect of Jδ on pluripotency gene. Following the protocol described before [24], the effect of our synthetic small molecules on the endogenous expression of iPSC factors was evaluated after 24 h incubation at 100 nM concentration. We chose the critical markers associated with pluripotency and cell fate conversion (Oct-3/4, Sox2, Nanog, Dppa4, Rex1, and Cdh1) as the candidate genes. First, we evaluated their effect independently. We observed, about 20- to 25-fold increase in the endogenous expression of Oct3/4 with both Sδ and Jδ. Then, we checked their effect in combination since they have different HDAC specificities. We used 50 nM of each of Sδ and Jδ in MEF. Interestingly, we observed together these two compounds strikingly improved the efficacy to induce Oct3/4 by about 1.6-times induced Oct-3/4 by about 40-fold (Fig. 2.5a, Bar Sδ + Jδ). Sδ and Jδ induced Nanog by about 20-fold, and 16-fold, respectively. Consistent with the pattern observed in Oct-3/4, Sδ and Jδ combination also showed improvement in the induction values (30-fold) of Nanog (Fig. 2.5b). Individual treatment of Sδ and Jδ induced the expression of Sox2 by about sixfold, whereas their combined treatment cause about tenfold increase in their induction values (Fig. 2.5c).

Rex1 is a critical pluripotency gene that is found in the undifferentiated cells and is silenced in somatic cells like MEF [33]. Interestingly, the combination of Sδ and Jδ induced the expression of Rex1 by about 100-fold in just 24 h (Fig. 2.5d, Bar Sδ +Jδ). Individual treatment of Sδ and Jδ also induced the Rex1 expression by about 60- to 70-fold. Cdh1 is a critical marker for mesenchymal epithelial transition (MET), an important rate-limiting step during the de-differentiation of the somatic genome [29]. Sδ and Jδ induced the expression of Cdh1 by about 25- to 30-fold when individually treated and in combination they induce the expression of Cdh1 by about 45-fold (Fig. 2.5e). In the case of Dppa4, only about two fold increase was observed with the individual and combined treatment of Sδ and Jδ (Fig. 2.5f). The acquisition of an epithelial fate during cellular reprogramming appears to be closely associated with the pluripotent state [34]. In this regard, dramatic induction of the core pluripotency genes including Cdh1 and Rex1 is just 24 h suggests that both Sδ and Jδ switch 'ON' the transcriptional machinery conferring to pluripotency. We reported previously that Sδ did not influence the expression of both c-Myc and Klf4; similarly, Jδ also did not have any effect on the induction of c-Myc and Klf4 when treated individually or in combination. The c-Myc and Klf4 pathways are different from that of the core pluripotency gene network [24] and our result is in accordance with this notion. Similar to Sδ, Jδ also displayed no cytotoxic effect on MEFs even at 10 μM, which implies that the cytotoxic effect is not correlated with its gene-inducing ability as shown in Fig. 2.8. Our results clearly suggest that Jδ retained the ability of Sδ to activate the pluripotency gene expression in mouse

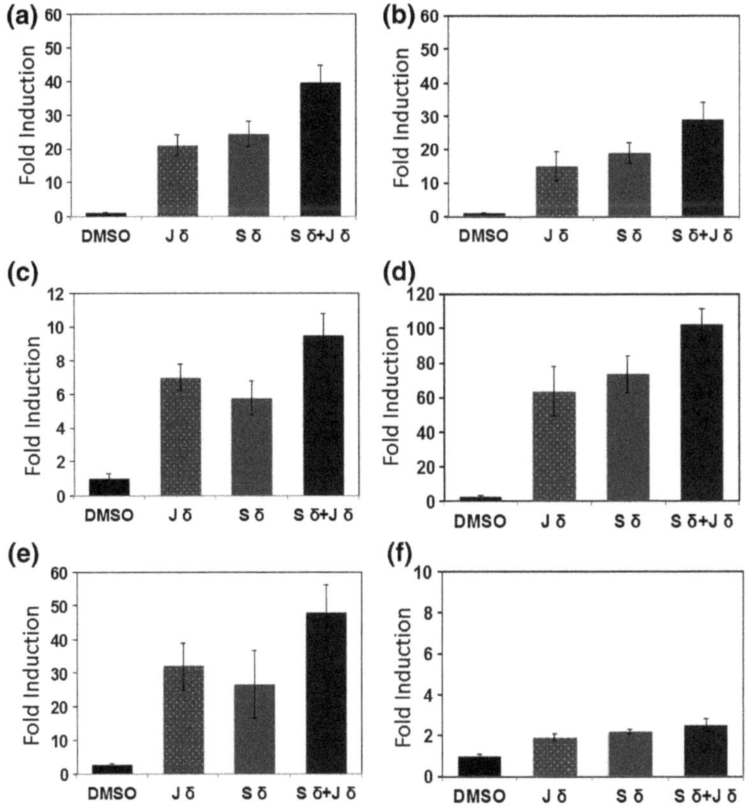

Fig. 2.5 Effect of Jδ on the endogenous expression of pluripotency genes. Expression levels of **a** *Oct3/4*, **b** *Nanog*, **c** *Sox2*, **d** *Rex1*, **e** *Cdh1*, and **f** *Dppa4* were determined by treating MEF with 100 nM of Sδ and Jδ either individually or together at 50 nM each, and with 0.1% DMSO as a control. Each bar represents the mean ± SD from 18 well plates

embryonic fibroblasts. Our results also suggest the possible need to inhibit HDAC8 for pluripotency gene expression as induction values of almost all pluripotency genes excluding Dppa4 was improved by combining Sδ with Jδ.

2.2.4 Chromatin Immunoprecipitation (ChIP)

In living cells, histone acetylation and deacetylation are catalyzed by histone acetyl transferase (HAT) and histone deacetylase (HDAC), respectively [33]. Histone acetylation is important in nucleosome assembly and chromatin folding as it controls the gene regulation [32]. Acetylation favors a loose chromatin structure by interfering with the interactions between nucleosomes and releasing histone tails

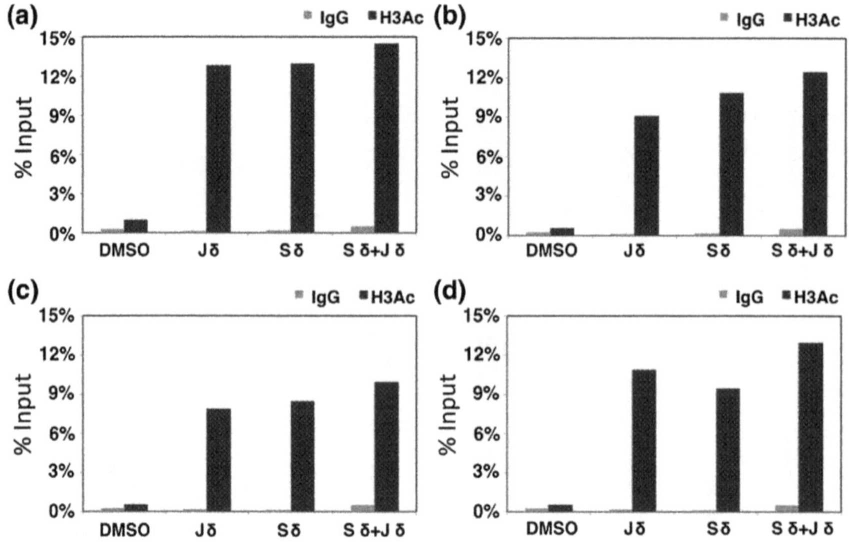

Fig. 2.6 Jδ retains the ability to hyper acetylate histone H3 in the promoter region of the core pluripotency genes. MEF were treated with Sδ and Jδ either individually or in combination as mentioned in the experimental section. After immunoprecipitation with H3ac antibody, the amount of promoter sequence of **a** *Oct3/4*, **b** *Nanog*, **c** *Cdh1*, **d** *Rex1* in the co-precipitated DNAs was determined by qPCR. 0.1% DMSO that was used as the control displayed no effect. Percent input was calculated by normalizing the data against input DNA, the enrichment with IgG antibody, and with internal control primers as mentioned previously [21]. The experiment included four treatments, each conducted in duplicate. Each bar represents the average of 12-well plates

from the linker DNA. Acetylation of lysine residues in histone H3 is commonly seen in genes that are being actively transcribed into RNA.

Therefore, MEF cells treated with Jδ were immunoprecipitated against acetylated histone H3 antibody to check the level of histone acetylation. We used Sδ and DMSO as positive and negative controls, respectively. We also employed the combination of Jδ and Sδ to show that their combination displayed better efficacy than the individual treatment. Our results indicate that Jδ triggered marked enhancement of the endogenous acetylation of histone H3 in the promoter region of Oct3/4, Nanog, Cdh1, and Rex1 (Fig. 2.6a–d). It is important to note here that the combination of Jδ with Sδ enhanced the enrichment in acetylated histones when compared to Sδ (Fig. 2.6a–d, Bars Sδ+Jδ). This result is consistent with the pattern observed in gene expression studies. Acetylation level of histone H3 in Sδ-treated MEFs was shown to be comparable and occasionally surpass than that in ES cells. Therefore, it is reasonable to suggest that our targeting small molecules (Jδ and Sδ) could epigenetically induce cellular reprogramming by switching "ON" some of the transcriptional gene network conferring to pluripotency.

2.2.5 Specific HDAC8 Inhibition

Next, we looked at HDAC8-regulated genes and their potential role in the developmental process. Haberland et al., revealed the central role of HDAC8 in skull morphogenesis through loss of function studies [28]. Their study in mice showed that HDAC8 specifically represses the aberrant expression of the homeobox transcription factors, Otx2 and Lhx1. The IC_{50} values shown before (Table 2.1) clearly suggested that Jδ is relatively specific against HDAC8 when compared to Sδ and TSA. Since Otx2 and Lhx1 were shown to be specifically regulated by HDAC8, we studied the effect of Sδ and Jδ on the expression of these two genes in MEF. DMSO alone did not have any effect on these genes (Fig. 2.7a, b, Bars DMSO). Jδ notably

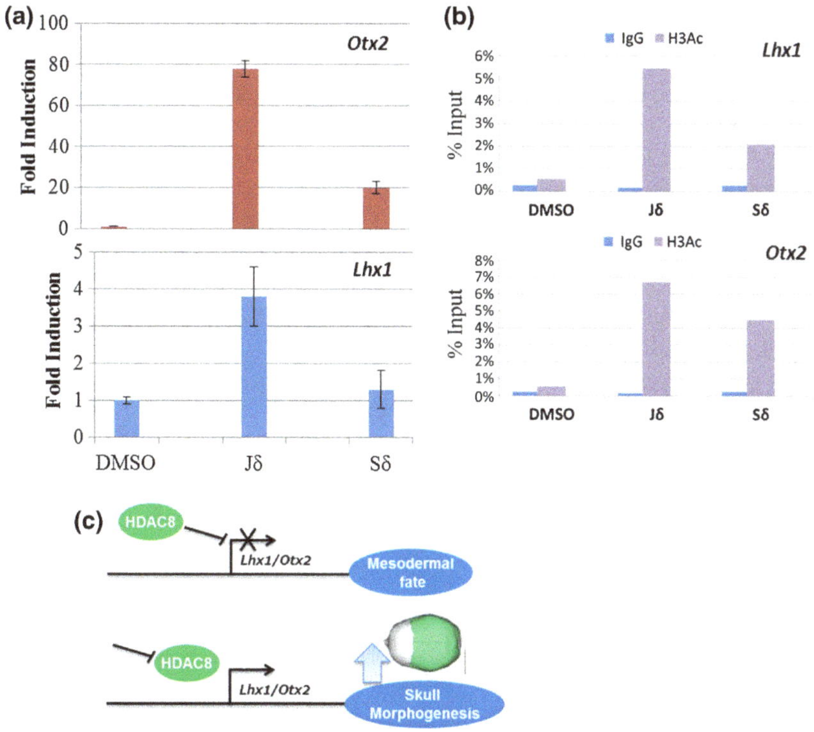

Fig. 2.7 **a** Effect of HDAC8-specific Jδ on the endogenous expression of skull morphogenesis associated *Lhx1* and *Otx2*. For the individual treatment of the effectors (Jδ and Sδ), the concentration used was 100 nM, which was based on the initial optimization studies. For the combined treatment, 50 nM of each effector was used as mentioned in the experimental section. **b** The amount of promoter sequence of *Lhx1* and *Otx2* in the co-precipitated DNA was determined by qPCR to calculate the percent input as mentioned in Fig. 2.6. **c** Schematic representation of the role of HDAC8 Inhibition in causing mesodermal fate by the inhibition of *Lhx1* and *Otx2* and inhibition of HDAC8 is known to promote neuroectodermal fate that is essential for skull morphogenesis (Upper panel) [28]. Inhibition of HDAC8 by Jδ results in induction of *Lhx1* and *Otx2*. Hence, Jδ could be improved to modulate cell fate (lower panel)

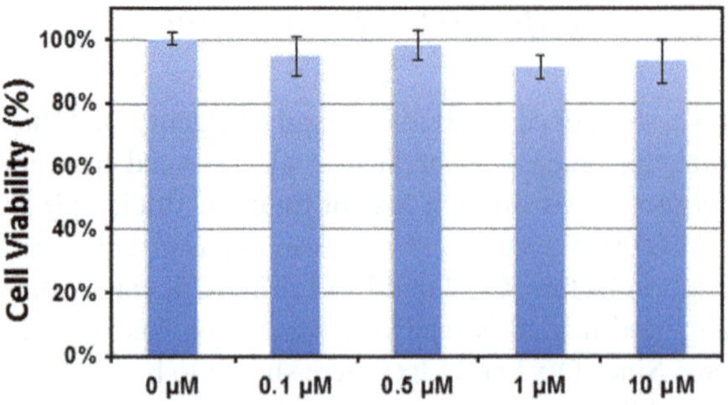

Fig. 2.8 Cytotoxicity assay of JAHA-δ. Cell viability of MEF was measured after 24 h treatment of the above effectors with various concentrations. Each bar represents mean ± SD from 12 wells

induced the expression of Otx2 and Lhx1 by about 74-fold and 2.6-fold, respectively (Fig. 2.7a, b, Bars Jδ). Sδ alone induced the endogenous expression of Otx2 by about 40-fold but had no effect on Lhx1 (Fig. 2.7a, b, Bars Sδ). Since Sδ is relatively specific against HDAC8, the difference in the acetylated histones was suggested to be the reason behind this differential gene expression. Hence, ChIP analysis was done to assess the acetylation status of histone H3 in the promoter region of Otx2 and Lhx1. Consistent with the pattern obtained with HDAC8 assay and gene expression studies, Jδ showed better enrichment of the acetylated histone H3 in Otx2 than that obtained with Sδ (Fig. 2.7b, Patterned Bar Jδ). A distinctive difference was observed in the case of Lhx1, where Jδ showed a notable increase in the level of acetylated histones. On the other hand, in comparison with DMSO, Sδ showed little or no increase in the level of acetylated histone H3 (Fig. 2.7b, Patterned Bar DMSO and Sδ). This result is consistent with our notion that differential gene expression is caused by the difference in the level of acetylated histones. Activation of HDAC8-regulated Otx2/Lhx1 shifts neural crest cells (NCC) to neuroectodermal, while their inhibition shifts NCC to mesodermal fate [28]. Hence, rapid activation of Otx2/Lhx1 by Jδ through HDAC8 inhibition is strikingly important as it could be used to change the cell fate as illustrated in Fig. 2.7c.

2.3 Discussion

We have developed a new class of small molecules, termed Sδ as targeted transcriptional activators that could differentially activate certain genes associated with pluripotency [22–24]. Microarray analysis suggested that Sδ could switch the transcriptional network from the fibroblast to the dedifferentiated state in just 24 h by rapidly overcoming the MET stage that is associated with initiation phase of cellular reprogramming. Unlike other small molecules currently employed to improve reprogramming efficiency, Sδ can be tailored to bind predetermined DNA sequences owing to their ease and flexibility of design and the presence of flexible sites for covalent attachment to other molecules [35]. Here, we explored the effect of modification in SAHA by introducing a new compound Jδ. In vitro enzymatic activity assay for an individual class I HDACs showed that Jδ, the synthesized derivative of Sδ, is relatively specific toward HDAC8 (Table 2.1). Jδ not only retained the ability of Sδ to epigenetically induce multiple pluripotency genes but also improved the efficacy of Sd, when incubated together (Figs. 2.5 and 2.6 Bars Sδ+Jδ). Since the combination of Jδ with Sδ increased the level of acetylated histones, the potential importance of HDAC8 inhibition in pluripotent gene activation is suggested. However, further studies are needed to validate this hypothesis. Nevertheless, it is clear that our programmable small molecules could selectively switch "ON" some of the transcriptional machinery conferring to pluripotency in just 24 h.

HDAC8 is known to be involved in several biological processes including centrosome cohesion and metabolic control of estrogen-related receptors [36]. One of the well-characterized functions of HDAC8 was their ability to epigenetically regulate the fate of neural cranial cells by repressing Otx2 and Lhx1 [28]. Jδ notably induced the endogenous expression of the HDAC8-specific Otx2 and Lhx1 via hyperacetylation of histone H3 in their promoter region. Otx2 is important in the development of the brain and the sense organs [37]. Recent studies have indicated Otx2 as the intrinsic determinant of the pluripotent state that antagonizes the ground state pluripotency to promote commitment to differentiation [38]. Lhx1 plays a vital role in early mesoderm formation and later in lateral mesoderm differentiation and neurogenesis [39]. Since these two genes are normally not expressed in somatic cells like MEF, their rapid activation by Jδ in somatic cells like MEF is biologically significant. Interestingly, SAHA alone did not induce the expression of these two genes [21]. Therefore, strategies to expand our tunable epigenetic-based genetic switches could create an epoch-making approach in cellular reprogramming as they may precisely coax the somatic cells into pluripotent stem cells and/or a totally new type of cells.

2.4 Summary

In summary, a new compound, Jδ, lacking the surface-recognition domain of SAHA, was synthesized. Through enzymatic activity assay and biological studies, it was demonstrated that the chemical modification of functional SAHA in SAHA-PIP was shown to alter its activity against HDAC8 but not against HDAC1. An additive effect was observed with the combination of HDAC8-specific Jδ with Sδ. Rapid activation of skull morphogenesis associated HDAC8 regulated Otx2 and Lhx1 were demonstrated for the first time to open up exciting opportunities in regenerative medicine. Recent progress in the development of programmable DNA-binding SAHA-PIPs suggests that it is possible to alter the epigenome in a site-specific manner through transcriptional activation. Synthetic SAHA-PIPs gain an advantage over other natural DNA-binding proteins as effective transcriptional activators because they can bind to the methylated DNA sequences and hence can disrupt the packed chromatin structure [40]. Small molecules should possess multifunctional properties to effectively mimic the natural transcription factors and reset the signaling or epigenetic pathways. In this regard, SAHA-PIP is shown here to not only have the flexibility to bind to predetermined DNA sequences but could also be altered to bind to different HDAC enzymes. Development of our targeted transcriptional activators could be useful in regenerative medicine. However, care should be taken in the design of these small molecules because chromatin remodeling is not an isolated event. Moreover, factors such as accessibility and genome-wide specificity need to be addressed to avoid side effects. Nevertheless, the development of small molecules with multiple, but specific functionalities open up new opportunities in the regulation of specific gene(s) of interest.

2.5 Experimental Section

2.5.1 Cell Culture and Treatment of PIP Conjugates in MEF

MEF cells in the sixth passage were trypsinized for 5 min at 37 °C and were resuspended in fresh DMEM medium to give a concentration of 2×10^5 cell/mL in a 30 mm dish culture plate. After overnight incubation for attachment, the medium was replaced with fresh DMEM (2 mL) containing each individual PIP conjugates (Sδ and Jδ) in 100 nM concentration. This concentration was found to be effective based on our previous reports and initial optimization studies. In the case of additive effect, both Sδ and Jδ were added together in 50 nM concentration each. DMSO (0.1%)-treated MEF was used for control. All the plates were incubated in a 5% CO_2-humidified atmosphere at 37 °C. MEF treated with both the effectors (Sδ and Jδ) were harvested at 24 h based on the previous studies [22–24].

C57BL/6 mouse embryonic fibroblasts (MEF) were purchased from the American Type Culture Collection (ATCC). MEF cells were cultured and maintained in Dulbecco's modified Eagle's medium (DMEM) supplemented with heat-inactivated fetal bovine serum (FBS) (15%), penicillin (100 IU/mL), and streptomycin (100 μg/mL) at 37 °C in a humidified atmosphere of CO_2 (5%) in air (95%).

2.5.2 Cytotoxicity Assay

Colorimetric assays using WST-8 (Dojindo, Kumamoto, Japan) were carried out in 96-well plates with various concentrations of the effectors DMSO, SAHA, and JAHA-δ.

2.5.3 *Biological Procedures*

HDAC enzyme activity was measured using the HDAC Fluorogenic Assay Kit supplied from BPS Bioscience. Experimental procedures were maintained in accordance with the supplier's instructions. Fluorescence measurements were obtained using a JASCO FP-6300 fluorimeter. Read sample using this fluorimeter was capable of excitation at a wavelength in the range of 350–380 nm and detection of emitted light in the range of 440–460 nm. Each plate was analyzed by plate repeat. Replicate experimental data from incubations with inhibitor were normalized to control IC_{50} determined by logistic regression.

2.5.4 *Quantification of Expression of Marker Genes in Mouse Embryonic Fibroblasts*

Total RNA was extracted from JAHA-δ, SAHA-δ, and DMSO-treated MEF using an RNeasy Mini Kit (Qiagen) and cDNA was synthesized with ReverTra Ace qPCR RT kit (Toyobo, Japan), in accordance with the manufacturer`s instructions. SYBR green real-time RT-PCR amplifications were carried out in triplicate with the protocol and conditions mentioned in THUNDERBIRD SYBR qPCR Mix (TOYOBO, Japan) on an ABI 7300 Real-Time Detection System (Applied Biosystems, USA) and were analyzed using a 7300 System SDS Software v1.3.0 (Applied Biosystems, USA). Melting curve analysis of amplification products was performed at the end of each PCR reaction to confirm that only one PCR product was amplified and detected. After normalization with housekeeping gene GAPDH, using the comparative cycle threshold (CT) method, the relative expression level of each gene was analyzed by considering the gene expression in DMSO-treated cells as 100%. Primer pairs used for the endogenous gene such as Oct-3/4, Nanog, Sox2, Klf4, Dppa4, Cdh1, Rex1 and c-Myc were reported before. The primers used for HDAC 8-specific genes such as Otx2 and Lhx1 were reported.

2.5.5 *Chromatin Immunoprecipitation (ChIP) Analysis*

ChIP assay was performed according to the protocol described in a previous report 7 and supplied kit manual. Antibodies for acetylated histone H3 and normal rabbit IgG were purchased from Upstate (Millipore) having the product name ChIPAb +acetyl-Histone H3. Compounds were treated in 100 nm concentrations to MEF and after 24 h incubation ChIP analysis was performed. ChIP-purified DNA fraction was then purified QIAquick PCR Purification Kit (Qiagen, USA) and analyzed with qRT-PCR using THUNDERBIRD SYBR qPCR Mix (TOYOBO, Japan). Primers used for the promoter region of mouse Oct3/4, Nanog, CDH1, and Rex1 for qPCR was reported [4]. For the control, input DNA that is collected after adding ChIP dilution buffer but before the addition of ChIP antibody was used. Enrichment fold is calculated by normalizing the data against input DNA and IgG as follows ΔCt [normalized ChIP] = (Ct [ChIP] − (Ct [Input] − Log2 (Input Dilution Factor))); ΔΔCt [ChIP/NS]=ΔCt [normalized ChIP] − ΔCt [normalized NS] NS indicated nonspecific antibody, here IgG. Finally, fold enrichment was calculated from 2-ΔΔCt to ensure the normalization of all the background signals.

2.5.6 PCR Protocol

PCR cycle:

Stage 1	50 °C for 2 min, 1 cycle
Stage 2	95 °C for 10 min, 1 cycle
Stage 3	95 °C for 15 s, 40 cycles
	60 °C for 1 min
Stage 4	95 °C for 15 s, 1 cycle
	60 °C for 30 s,
	95 °C for 15 s.

References

1. Wähnert U, Zimmer O, Luck G, Pitra O (1975) (dA-dT) Dependent inactivation of the DNA template properties by interaction with netropsin and distamycin A. Nucleic Acids Res 2:391–404
2. Kopka ML, Yoon C, Goodsell D, Pjura P, Dickerson RE (1985) Binding of an antitumor drug to DNA, netropsin and C-G-C-G-A-A-T-T-BrC-G-C-G. Mol Biol 183:553–563
3. Buchmueller KL, Staples AM, Howard CM, Horick SM, Uthe PB, Le NM, Cox KK, Nguyen B, Pacheco KA, Wilson WD, Lee M (2005) Extending the language of DNA molecular recognition by polyamides: unexpected influence of imidazole and pyrrole arrangement on binding affinity and specificity. J Am Chem Soc 127:742–750. https://doi.org/10.1021/ja044359p
4. Gottesfeld JM, Neely L, Trauger JW, Baird EE, Dervan PB (1997) Regulation of gene expression by small molecules. Nature 387:202–205
5. Bando T, Sugiyama H (2006) Synthesis and biological properties of sequence-specific DNA-alkylating pyrrole-imidazole polyamides. Acc Chem Res 39:935–944
6. Nishijima S, Shinohara K, Bando T, Minoshima M, Kashiwazaki G, Sugiyama H (2010) Cell permeability of Py-Im-polyamide-fluorescein conjugates: influence of molecular size and Py/Im content. Bioorg Med Chem 18:978–983. https://doi.org/10.1016/jbmc200907018
7. Tao ZF, Fujiwara T, Saito I, Sugiyama H (1999) Rational design of sequence-specific DNA alkylating agents based on duocarmycin A and pyrrole–imidazole hairpin polyamides. J Am Chem Soc 121:4961–4967. https://doi.org/10.1021/ja983398w

8. Tao ZF, Saito I, Sugiyama H (2000) Highly cooperative DNA Dialkylation by the homodimer of imidazole–pyrrole diamide–CPI conjugate with vinyl linker. J Am Chem Soc 122:1602–1608. https://doi.org/10.1021/ja9926212
9. Kashiwazaki G, Bando T, Yoshidome T, Masui S, Takagaki T, Hashiya K, Pandian GN, Yasuoka J, Akiyoshi K, Sugiyama H (2012) Synthesis and biological properties of highly sequence-specific-alkylating N-methylpyrrole-N-methylimidazole polyamide conjugates. J Med Chem 55:2057–2066. https://doi.org/10.1021/jm201225z
10. Takagaki T, Bando T, Sugiyama H (2012) Synthesis of pyrrole-imidazole polyamide seco-1-chloromethyl-5-hydroxy-1,2-dihydro-3 h-benz[e]indole conjugates with a vinyl linker recognizing a 7 bp DNA sequence. J Am Chem Soc 134:13074–13081. https://doi.org/10.1021/ja3044294
11. Best TP, Edelson BS, Nickols NG, Dervan PB (2003) Nuclear localization of pyrrole-imidazole polyamide-fluorescein conjugates in cell culture. Proc Natl Acad Sci USA 100:12063–12068. https://doi.org/10.1073/pnas.2035074100
12. Fechter EJ, Olenyuk B, Dervan PB (2005) Sequence-specific fluorescence detection of DNA by Polyamide–Thiazole orange conjugates. J Am Chem Soc 127:16685–16691. https://doi.org/10.1021/ja054650k
13. Correa BJ, Canzio D, Kahane AL, Reddy PM, Bruice TC (2006) DNA sequence recognition by Hoechst 33258 conjugates of hairpin pyrrole/imidazole polyamides. Bioorg Med Chem Lett 16:3745–3750. https://doi.org/10.1016/j.bmcl.2006.04.047
14. Puckett JW, Muzikar KA, Tietjen J, Warren CL, Ansari AZ, Dervan PB (2007) Quantitative microarray profiling of DNA-binding molecules. J Am Chem Soc 129:12310–12319. https://doi.org/10.1021/ja0744899
15. Vaijayanthi T, Bando T, Pandian GN, Sugiyama H (2012) Progress and prospects of pyrrole-imidazole polyamide-fluorophore conjugates as sequence-selective DNA probes. Chem BioChem 13:2170–2185. https://doi.org/10.1002/cbic.201200451
16. Vaijayanthi T, Bando T, Hashiya K, Pandian GN, Sugiyama H (2013) Design of a new fluorescent probe: pyrrole/imidazole hairpin polyamides with pyrene conjugation at their γ-turn. Bioorg Med Chem 21:852–855. https://doi.org/10.1016/j.bmc.2012.12.018
17. Shi Y, Do JT, Desponts C, Hahm HS, Schöler HR, Ding S (2008) A combined chemical and genetic approach for the generation of induced pluripotent stem cells. Cell Stem Cell 2:525–528. https://doi.org/10.1016/j.stem.2008.05.011
18. Huangfu D, Maehr R, Guo W, Eijkelenboom A, Snitow M, Chen AE, Melton DA (2008) Induction of pluripotent stem cells by defined factors is greatly improved by small-molecule compounds. Nat Biotechnol 26:795–797. https://doi.org/10.1038/nbt1418
19. Mikkelsen TS, Hanna J, Zhang X, Ku M, Wernig M, Schorderet P, Bernstein BE, Jaenisch R, Lander ES, Meissner A (2008) Dissecting direct reprogramming through integrative genomic analysis. Nature 454:49–55. https://doi.org/10.1038/nature07056
20. Ohtsuki A, Kimura MT, Minoshima M, Suzuki T, Ikeda M, Bando T, Nagase H, Shinohara K, Sugiyama H (2009) Synthesis and properties of PI polyamide–SAHA conjugate. Tetrahedron Lett 50:7288–7292. https://doi.org/10.1016/j.tetlet.2009.10.034
21. Pandian GN, Sugiyama H (2012) Programmable genetic switches to control transcriptional machinery of pluripotency. Biotechnol J 7:798–809. https://doi.org/10.1002/biot.201100361
22. Pandian GN et al (2011) Synthetic small molecules for epigenetic activation of pluripotency genes in mouse embryonic fibroblasts. Chem BioChem 12:2822–2828. https://doi.org/10.1002/cbic.201100597
23. Pandian GN, Ohtsuki A, Bando T, Sato S, Hashiya K, Sugiyama H (2012) Development of programmable small DNA-binding molecules with epigenetic activity for induction of core pluripotency genes. Bioorg Med Chem 20:2656–2660. https://doi.org/10.1016/j.bmc.2012.02.032
24. Pandian GN, Nakano Y, Sato S, Morinaga H, Bando T, Nagase H, Sugiyama H (2012) A synthetic small molecule for rapid induction of multiple pluripotency genes in mouse embryonic fibroblasts. Sci Rep 2:544. https://doi.org/10.1038/srep00544

25. Pontiki E, Lintina DH (2012) Histone deacetylase inhibitors (HDACIs). Structure—activity relationships: history and new QSAR perspectives. Med Res Rev 32:1–165. https://doi.org/10.1002/med.20200
26. Finnin MS, Donigian JR, Cohen A, Richon VM, Rifkind RA, Marks PA, Breslow RN, Pavletich P (1999) Structures of a histone deacetylase homologue bound to the TSA and SAHA inhibitors. Nature 401:188–193. https://doi.org/10.1038/43710
27. Spencer J, Amin J, Wang M, Packham G, Alwi SS, Tizzard GJ, Coles SJ, Paranal RM, Bradner JE, Heightman TD (2011) ACS Med Chem Lett 2:358–362. https://doi.org/10.1021/ml100295v
28. Haberland M, Mokalled MH, Montgomery RL, Olson EN (2009) Epigenetic control of skull morphogenesis by histone deacetylase 8. Gene Dev 23:1625–1630. https://doi.org/10.1101/gad.1809209
29. Butler KV, Kalin J, Brochier C, Vistoli G, Langley B, Kozikowski AP (2010) Rational design and simple chemistry yield a superior, neuroprotective HDAC6 inhibitor, tubastatin A. J Am Chem Soc 132:10842–10846. https://doi.org/10.1021/ja102758v
30. Kidder BL, Palmer S (2012) HDAC1 regulates pluripotency and lineage specific transcriptional networks in embryonic and trophoblast stem cells. Nucleic Acids Res 40:2925–2939. https://doi.org/10.1093/nar/gkr1151
31. Vega RB, Matsuda K, Oh J, Barbosa AC, Yang X, Meadows E, McAnally J, Pomajzl C, Shelton JM, Richardson JA, Karsenty G, Olson EN (2004) Histone deacetylase 4 controls chondrocyte hypertrophy during skeletogenesis. Cell 119:555–566. https://doi.org/10.1016/j.cell.2004.10.024
32. Pandian GN, Sugiyama H (2012) Strategies to modulate heritable epigenetic defects in cellular machinery: lessons from nature. Pharmaceuticals 6:1–24. https://doi.org/10.3390/ph6010001
33. Wang J, Rao S, Chu J, Shen X, Levasseur DN, Theunissen TW, Orkin SH (2006) A protein interaction network for pluripotency of embryonic stem cells. Nature 444:364–368. https://doi.org/10.1038/nature05284
34. Li R, Liang J, Ni S, Zhou T, Qing X, Li H, He W, Chen J, Li F, Zhuang Q, Qin B, Xu J, Li W, Yang J, Gan Y, Qin D, Feng S, Song H, Yang D, Zhang B, Zeng L, Lai L, Esteban MA, Pei D (2010) A mesenchymal-to-epithelial transition initiates and is required for the nuclear reprogramming of mouse fibroblasts. Cell Stem Cell 7:51–63. https://doi.org/10.1016/j.stem.2010.04.014
35. Jacobs CS, Dervan PB (2009) Modifications at the C-terminus to improve pyrrole-imidazole polyamide activity in cell culture. J Med Chem 52:7380–7388. https://doi.org/10.1021/jm900256f
36. Wilson BJ, Tremblay AM, Deblois G, Sylvain-Drolet G, Giguère V (2010) An acetylation switch modulates the transcriptional activity of estrogen-related receptor alpha. Mol Endocrinol 24:1349–1358. https://doi.org/10.1210/me.2009-0441
37. Puelles E, Annino A, Tuorto F, Usiello A, Acampora D, Czerny T, Brodski C, Ang SL, Wurst W, Simeone A (2004) Otx2 regulates the extent, identity and fate of neuronal progenitor domains in the ventral midbrain. Development 131:2037–2048. https://doi.org/10.1242/dev.01107
38. Acampora D, Di Giovannantonio LG, Simeone A (2013) Otx2 is an intrinsic determinant of the embryonic stem cell state and is required for transition to a stable epiblast stem cell condition. Development 140:43–55. https://doi.org/10.1242/dev.085290
39. Shawlot W, Wakamiya M, Kwan KM, Kania A, Jessell TM, Behringer RR (1999) Lim1 is required in both primitive streak-derived tissues and visceral endoderm for head formation in the mouse. Development 126:4925–4932
40. Minoshima M, Bando T, Sasaki S, Fujimoto J, Sugiyama H (2008) Pyrrole-imidazole hairpin polyamides with high affinity at 5′-CGCG-3′ DNA sequence; influence of cytosine methylation on binding. Nucleic Acids Res 36:2889–2894. https://doi.org/10.1093/nar/gkn116

Chapter 3
Chemical Modification of a Synthetic Small Molecule Boosts Its Biological Efficacy Against Pluripotency Genes in Mouse Fibroblast

Abstract Our synthetic transcriptional activator SAHA-PIP, Sδ (described here as 1), encompassing both sequence-specific pyrrole–imidazole polyamides (PIPs) and an epigenetic activator (SAHA) was shown to induce the endogenous expression of core pluripotency genes in mouse embryonic fibroblasts (MEFs). However, the expression levels of pluripotency genes by **1** in MEFs were relatively lesser than that in mouse embryonic stem (ES) cells. Here, in this chapter, we carried out studies to improve the efficacy of **1** and show that the biological activity of **1** got significantly ($P \leq 0.05$) improved against the core pluripotency genes after the incorporation of an isophthalic acid (IPA) in its C-terminus. The resultant IPA conjugate **2** dramatically induced Oct-3/4 to demonstrate a new chemical strategy for developing PIP conjugates as next-generation genetic switches.

Keywords Biological activity · Cellular uptake · Gene expression Polyamides · Transcription

3.1 Introduction

A recent study showed an artificial induction of pluripotency in mouse somatic cells using a combination of seven small molecules [1]. Since this transgene-free approach enhances the clinical prospects of induced pluripotent stem (iPS) cell technology, screening and identification of small molecules that could influence the pluripotency genes are now in rising demand [2a]. Several chemical modulators of epigenetic enzymes and/or signaling pathway factors have shown success in enhancing the somatic cell reprogramming [2b]. However, the lack of selectivity and requirement of many small molecules is a major concern. In this regard, we previously conjugated the histone deacetylase inhibitor SAHA (suberoylanilide hydroxamic acid) with hairpin pyrrole–imidazole polyamide (PIPs) to create a new class of small molecules called SAHA-PIP [3]. PIPs are sequence-specific DNA minor groove-binding small molecules having the binding ability similar to natural transcriptional binding proteins [4, 5]. Sequence selectivity of PIPs is programmed

A. Saha, *Molecular Recognition of DNA Double Helix*, Springer Theses,
https://doi.org/10.1007/978-981-10-8746-2_3

by the side by side stacked ring pairing, where I/P distinguishes G.C from C.G and P/P pair with either A.T or T.A [6]. Hence, distinctive PIPs may direct SAHA to their match sequences and induce differential transcriptional activation [7, 8]. Screening studies carried out for evaluating the effect of 32 distinct SAHA-PIPs on pluripotency genes in mouse embryonic fibroblasts (MEFs) revealed that certain SAHA-PIPs could differentially induce Oct-3/4, Sox2, Klf4, and c-Myc [8]. In particular, SAHA-PIP 1 (Fig. 3.1), targeting 5′-WGWWC-3′ (W = A/T) sequence induced core pluripotency genes in just 24 h with 100 nM treatment [9–11]. Extensive analysis of the microarray data on genome-wide gene expression in MEFs revealed that 1 significantly induce Oct-3/4 pathway genes that are related to embryonic stem cell pluripotency [10]. None of the top ten significantly enhanced pathways in SAHA-treated MEF were related to embryonic stem cell pluripotency. Although, 1 could notably upregulate pluripotency genes in MEFs, the expression levels were still lower when compared to that in ES cells.

A general hypothesis for PIP biological activity relied on its cell uptake and competing for efflux pathway can influence its accumulation in cell nucleus and hinder its effect on endogenous gene expression [12]. Therefore, a decrease in efflux could increase the biological activity of PIP [13]. Hence, in this study we first employed the carbohydrate-solubilizing agent 2-hydroxypropyl-β-cyclodextrin (HpβCD) in cell culture medium. Since PIPs are known to form measurable particle size or aggregates under biologically relevant conditions [13], we checked the ability of HpβCD to reduce the aggregation of 1. HpβCD increased the solubility of 1 under in vitro conditions, but the biological efficacy of 1 was only marginally improved.

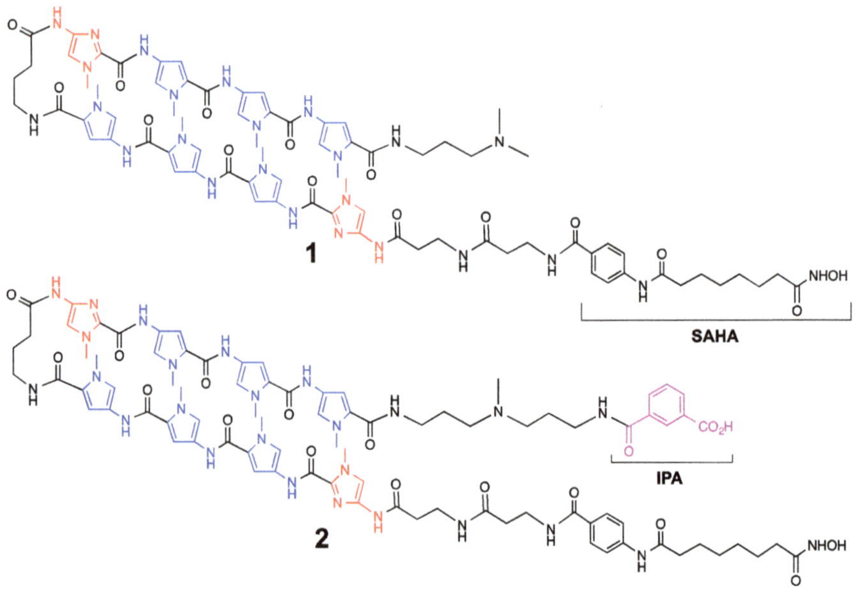

Fig. 3.1 Chemical structures of 1 and its C-terminus-modified analog as 2

A varied structural motif change at the C-terminus of a PIP has been studied by Dervan and co-workers and among many choices, an isophthalic acid (IPA) incorporation notably enhanced its solubility [12]. Therefore, we incorporated IPA in the C-terminus of 1 to synthesize a new compound called 2 (Fig. 3.1). Through real-time PCR studies, we demonstrate here for the first time that the biological activity of a SAHA-PIP could be significantly enhanced with the incorporation of an IPA moiety in its C-terminus.

3.2 Results and Discussion

3.2.1 *Solubility of 1 in the Presence of* Hp*β*CD

Increasing the solubility of 1 could improve its biological efficacy as cellular uptake and nuclear localization often hinder the PIP-mediated endogenous gene regulation. Employment of 2-hydroxypropyl-*β*-cyclodextrin (Hp*β*CD) is known to increase the solubility of PIP, however, its effect on SAHA-PIP like 1 is unknown. In order to determine the effect of Hp*β*CD on 1, first we checked its solubility in the presence or absence of cyclodextrin. By measuring the HPLC peak area detected at 254 nm, the macroscopic solubility of the respective PIP solutions (same concentration) were investigated in the presence of 0–50 mM Hp*β*CD. For comparison of the solubility, the peak area of the same PIPs dissolved in DMSO was assigned as 100%.

For each PIP, two kinds of solubility were measured; one in the presence of Hp*β*CD (+) and another in its absence (−). Conjugate 3 containing an acetyl cap at the N-terminus showed enhancement in solubility in the presence of CD (+). On the other hand, in the absence of CD (−), the solubility became lower. Interestingly, conjugate 1 having *β*-Ala-*β*-Ala-SAHA at the N-terminus showed reduced solubility in absence of CD (−); however its solubility got dramatically increased in the presence of CD (+). This result suggests that aggregation or precipitation occur in both 1 and 3 under biologically relevant conditions. Furthermore, this result implies that the presence of *ββ*SAHA at the N-terminus could increase the aggregation propensity in SAHA-PIPs (Fig. 3.2).

For checking the solubility of 1 using Hp*β*CD, we choose another compound 3, which is a precursor of 1 without having SAHA part. For each PIP (1 and 3), two types of solubility were checked. The one is in the presence of Hp*β*CD (+) and another one is in its absence (−). Conjugate 3 containing an acetyl cap at the N-terminus showed enhancement in solubility in the presence of CD (+). On the other hand, in the absence of CD (−), the solubility became lower. Interestingly, conjugate 1 having *β*-Ala-*β*-Ala-SAHA at the N-terminus showed reduced solubility in absence of CD (−); however, its solubility got dramatically increased in the presence of CD (+). This result suggests that aggregation or precipitation occur in both 1 and 3 under biologically relevant conditions. Furthermore, this result implies that the presence of *ββ*SAHA at the N-terminus could increase the aggregation propensity in SAHA-PIPs.

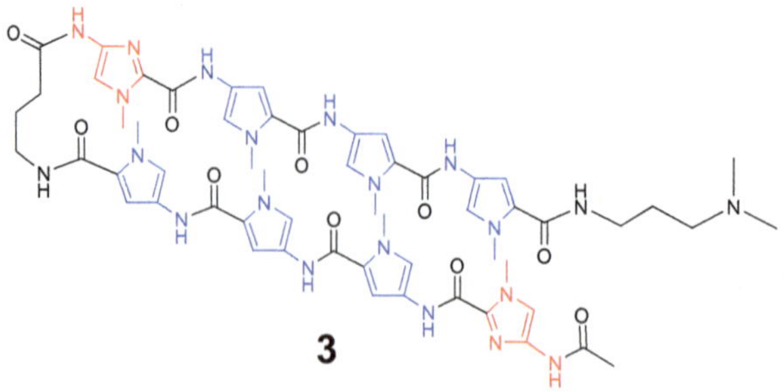

3

solubility
1 CD (-) : 3-4%
1 CD (+) : 31-36%
3 CD (-) : 25-41%
3 CD (+) : 51-54%

Fig. 3.2 Calculated solubility of polyamide 1 and 3 in 0.1% DMSO/PBS containing 0–50 mM Hp*β*CD at room temperature. Resultant solubility was determined by HPLC peak area at 254 nm detection. 100% solubility was calculated from the peak area of the respective polyamide dissolved only in DMSO

3.2.2 Effect of Hp*β*CD *in MEF Treated by 1*

Next, we thought if the barrier of solubility could be overcome by employing Hp*β*CD in cell culture. Since the endogenous mRNA levels of the target genes can be inferred as a biological readout of the effectors, quantitative real-time polymerase chain reaction (qRT-PCR) was used to monitor the cellular uptake of PIPs. As compound 1 can induce Oct-3/4 pathway genes, thus those selected genes were used. Relying on the fact, we carried out screening studies to evaluate the effect of CD (5 or 50 mM) on the endogenous expression of the core pluripotency genes. While 1 notably induced the endogenous expression of the core pluripotency genes as reported before,10 CD alone did not influence any of the genes under study (Fig. 3.3). Treatment of 1 dissolved in either 5 or 50 mM of Hp*β*CD showed only a marginal (a maximum of about 1.5-fold) increase in the endogenous expression of the core pluripotency genes (Oct-3/4, Nanog, Cdh1, Rex1, Sox2, and Dppa4) (Fig. 3.3). Statistical analyses verified that this marginal increase observed with the treatment of 1 with CD (5 or 50 mM) is not significant. Therefore, CD could improve the solubility of 1 under in vitro conditions but as a vehicle it could not significantly enhance its biological activity. Thus, an alternative strategy is required to improve the biological efficacy of 1.

3.2.3 Synthesis of Isophthalic Acid (IPA) Tail and 2

Previous gene regulation studies in cell culture relied on the results obtained with fluorescent-labeled PIPs as a cell-permeable small molecule [14]. In recent years, a non-fluorescent agent like isophthalic acid (IPA) at the C-terminus of a polyamide has substituted the fluorescent tag and rivaled the activity of the original FITC-labeed polyamides [12, 15]. Encouraged with these previous reports,

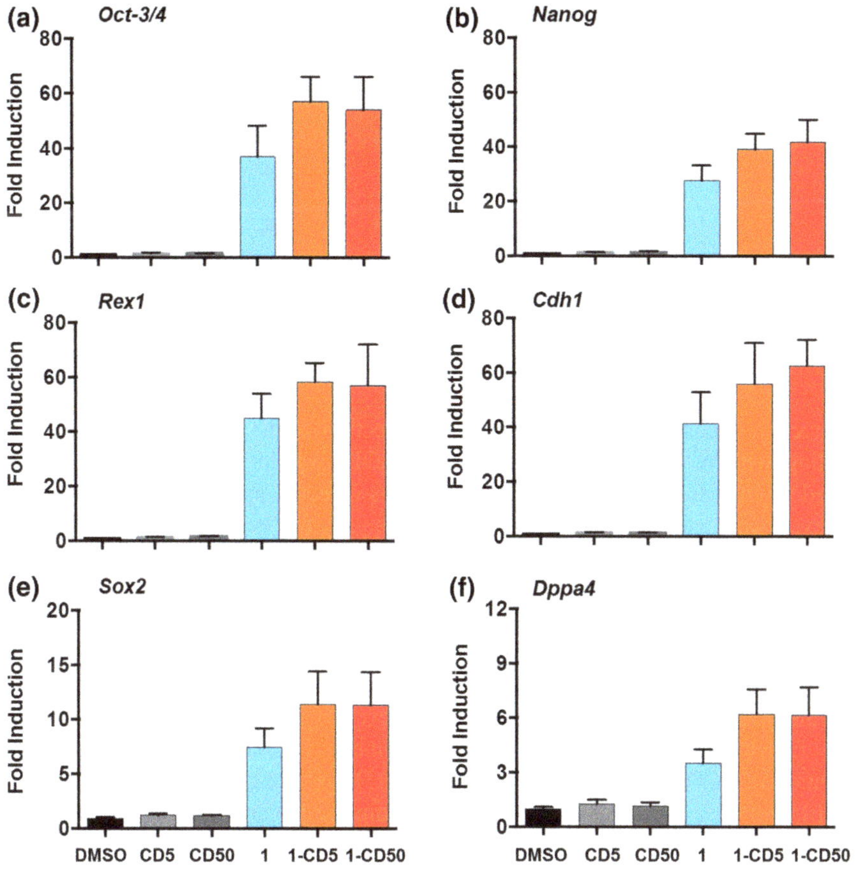

Fig. 3.3 Effect of HpβCD (5 and 50 mM) on 1 on the endogenous expression of pluripotency genes. Expression levels of **a** Oct3/4, **b** Nanog, **c** Rex1, **d** Cdh1, **e** Sox2, and **f** Dppa4 were determined by treating MEF with 100 nM of 1 to each sample, and with 0.1% DMSO, 5 and 50 mM of CD as a control experiment. Each bar represents the mean ± SD from 12-well plates

we analyzed the biological activity of 1 after the incorporation of an IPA moiety in its C-terminus. However, the synthesis scheme of IPA conjugated SAHA-PIP is difficult with the existing synthetic route. Hence, a milder approach to achieve such compound in good yield is proposed. Herein, we described the synthesis of IPA conjugated 1, termed as 2. We incorporated an isophthalic acid (IPA) at the C-terminus of 1 by first making an IPA-tail (Scheme 3.1) and then coupled it with COOH group present at the C-terminus of the PIP as shown in Scheme 3.2.

The synthesis of IPA-tail was started with the monoprotection of isophthalic acid (IPA) [16] using benzyl bromide and then we tried monoprotection of 3, 3-diamine N-methyldipropylamine (triamine). In a previous protocol, mono Boc-protected triamine was used to synthesize the IPA-tail [17]. However, the synthesis is difficult as while we tried the mono Boc protection of the triamine, a large amount of

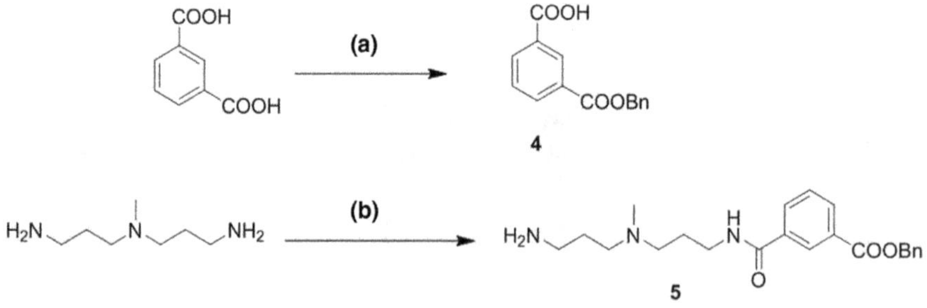

Scheme 3.1 Synthesis of C-terminus IPA-tail. Reagents and conditions: **a** BnBr, NEt3, MeOH/H2O, DMF, reflux; **b** 4, PyBOP, DMF, rt

diprotection occurred. Then without monoprotection of the triamine, we tried the coupling with monoprotected IPA using PyBOP. Within 1 h, we obtained our desired product in 80% yield while the rest was detected by decoupling product.

Next, we synthesized the polyamide sequence (Boc-β-β-Im-Py-Py-Py-γ-Im-Py-Py-Py-COOH) 6, using Fmoc solid phase chemistry by PSSM-8 peptide synthesizer (Shimadzu, Kyoto) with computer-assisted operating system. Already one pyrrole-loaded oxime resin 45 mg (ca. 0.2 mmol/g, 200–400 mesh) was used to start the synthesis in the solid-phase machine. Fmoc unit (such as Fmoc-Py-COOH, Fmoc-Im-COOH, or Fmoc-β-Ala-OH) (0.20 mmol) used in each step, were dissolved in NMP. The condition in each cycle of solid phase synthesis was as follows: deblocking (twice) with 20% piperidine/DMF (0.6 mL) for 4 min, pre-activation of –COOH group for 2 min with HCTU (88 mg, 0.21 mmol) in DMF (1 mL), and 10% DIEA/DMF (0.4 mL), coupling for 1 h and washing with DMF. After SPPS, the polyamide was detached from the resin using 2 M NaOH (aq.) in dioxane at 55 °C for 3 h (Scheme 3.2). The reaction mixture was filtered and washed with Et_2O, to yield a crude yellow powder. Without further purification, the precursor 6 (20 mg) was dissolved in DMF (0.5 mL) and coupled with pre-assembled IPA-tail (5) using PyBOP and DIEA. This allowed the formation of precursor 7 containing C-terminus IPA-tail. The reaction was quenched by adding water and then washed with Et_2O several times to achieve powder. The crude 7 without further purification was reduced using H_2/Pd-C and then Boc deprotection was done at the N-terminus using TFA. Excess TFA was removed by a vacuum pump and the crude reaction mixture was washed with Et_2O, which yielded a crude gray powder. In the next step, the polyamide having free amine at N-terminus was coupled with HOOC-SAHA-(Ome) using PyBOP and DIEA. This reaction allowed the formation of methylester of 2. In the last step, methyl ester of SAHA got converted to the active NH_2OH and provided the final product 2 after the reaction in aqueous solution of 50% (v/v) NH_2OH (0.15 mL).

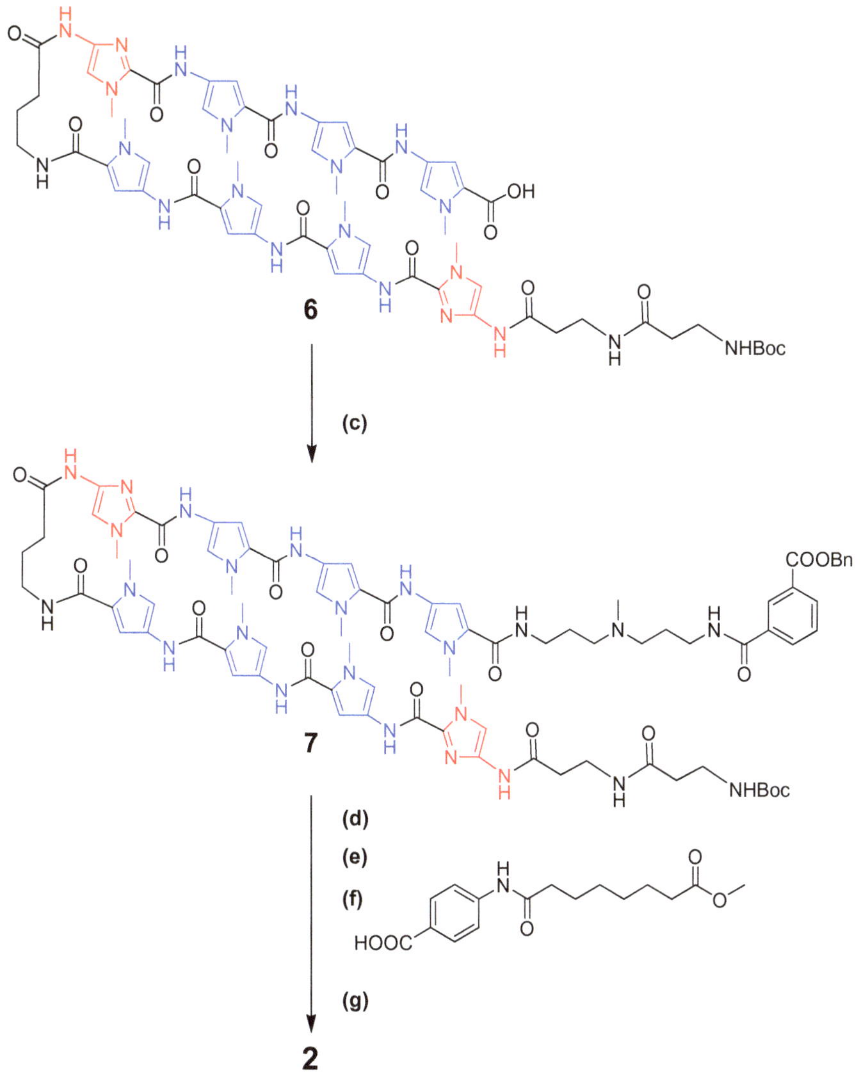

Scheme 3.2 Design and synthesis of isophthalic acid (IPA)-conjugated SAHA-PIP 2. Reagents and conditions: **c** 5, PyBOP, DIEA, DMF, rt; **d** Pd-C, H_2 (g), MeOH, rt; **e** 10% DCM in TFA, rt; **f** PyBOP, DIEA, DMF, rt; **g** 50% (v/v) NH_2OH, NMP, rt

3.2.4 *The Effect of Compound 2 in MEF*

MEF cells were treated with the purified 1 and 2 at the concentration of 100 nM. After 24 h of incubation, the effect of 1 and 2 on the endogenous expression of the core pluripotency genes was studied. The compound 2 induced the endogenous expression

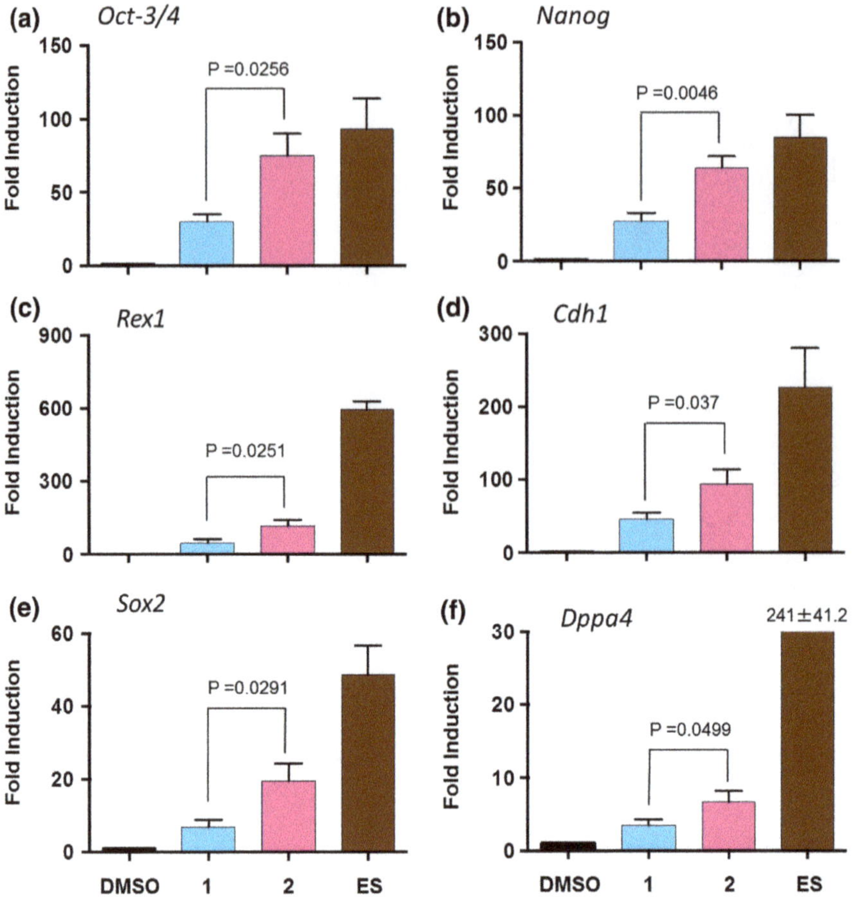

Fig. 3.4 Effect of 2 on the endogenous expression of pluripotency genes. Expression levels of **a** Oct3/4, **b** Nanog, **c** Rex1, **d** Cdh1, **e** Sox2, and **f** Dppa4 were determined by treating MEF with 100 nM of 1 and 2 individually, and with 0.1% DMSO as a control. ES cells were used as positive control. Each bar represents the mean ± SD from 12 well plates. Statistical significance was determined by t-test and p-values less than 0.05 is considered to be significant

of Oct-3/4 and Nanog by about 80-fold and 60-fold, respectively (Fig. 3.4a, b). These induction values were significantly ($P < 0.05$) higher than that observed in MEFs treated with 1 (Fig. 3.4a, b). Likewise, in 2-treated MEFs, about 100-fold increase in the endogenous expression of Rex1, which is a critical pluripotency gene typically silenced in somatic cells like MEFs was observed (Fig. 3.4c). Also, 2 notably induced Cdh1, a critical mesenchymal–epithelial transition (MET) marker by about 100-fold (Fig. 3.4d). The critical pluripotency genes like Sox2 and Dppa4 also got notably upregulated in 2-treated MEFs (Fig. 3.4d, e). It is important to note here that the expression levels of all the core pluripotency genes in 2-treated MEFs were significantly ($P < 0.05$) higher than that in 1-treated MEFs.

Table 3.1 Statistical analysis of SAHA-PIP 2 versus ES cells shown in Fig. 3.4

Genes	2 versus ES
Oct3/4	0.3
Nanog	0.0240
Rex1	0.0004
Cdh1	0.0358
Sox2	0.0102
Dppa4	0.0151

The expression levels of pluripotency genes clearly indicated that the incorporation of IPA at the C-terminus of 1 could significantly ($P < 0.05$) boost its biological efficacy. Interestingly, the expression level of Oct-3/4 in 2-treated MEFs was not significantly different from that observed in ES cells (Table 3.1).

Therefore, taken together, IPA incorporation could contribute as a vital troubleshooting strategy while designing PIPs capable of enhanced DNA recognition.

In recent years, many groups including our group have extensively studied the issue of PIP solubility. Consequently, many new developmental solutions have been developed to enhance the cellular uptake including pyrrole/imidazole content [18], C-terminus modification with isophthalic acid [11], and the presence of β-aryl turn at the γ-aminobutyric acid turn of a hairpin polyamide [19]. In particular, designing PIPs with more number of DNA base pair recognition ability could be hindered due to their poor cellular uptake. SAHA-PIP 1 having 6 base pair DNA recognition sequence can switch "ON" the silent pluripotency gene network in MEFs but with lower induction values when compared to that observed in ES cells. The efflux of PIP could be a limiting factor or a sufficient number of PIPs could not be reaching the target region of the gene(s) to activate transcription. Since the incorporation of IPA is known to increase the efficacy of PIPs [12], we synthesized the IPA conjugated SAHA-PIP 2 and tested its biological efficacy using qRT-PCR. In accordance with our notion, the biological efficacy of 2 was significantly higher than 1 in almost all tested genes. Recently certain SAHA-PIPs were shown to distinctively activate developmental genes in human dermal fibroblasts. Longer PIPs would be more specific to target sequences, however, the cellular uptake of such PIPs could hamper its use in gene regulation. The novel synthetic route described in this study to achieve PIP-IPA would be helpful for the future development of PIPs, which will be aiming at the selective induction of a therapeutically important or cell-fate-specifying target gene. We propose that such a challenging feat could be achieved through chemical modification of PIP such as IPA incorporation.

3.3 Conclusion

In summary, we carried out studies to enhance the efficacy of 1 and focused on improving the cellular uptake of 1 in cell culture medium. We first used the carbohydrate formulating agent cyclodextrin to prevent the aggregation of the polyamide, which in turn could increase its biological activity. The solubility of 1 could indeed be improved in the presence of cyclodextrin; however, the effect of 1 did not have a significant impact on the endogenous expression of its target pluripotency genes. In contrast, the alternative approach of modifying the chemical structure of 1 to synthesize the isophthalic acid (IPA) containing SAHA-PIP 2 had a significant impact as all the core pluripotency genes could be dramatically induced by 2. In our new synthetic procedure, we also overcome the low yield of the existing protocol and generate a PIP-IPA conjugate with high purity. Introduction of IPA in SAHA-PIP significantly improved its potency in pluripotency gene expression. Therefore, such chemically modified SAHA-PIPs could find its use in cellular reprogramming that requires long-term treatment. Also, the lesser number and concentration of effectors could help us to overcome the challenges associated with the solubility of PIP conjugates and its formulation. The amenability of IPA conjugation at the C-terminus of SAHA-PIP also open up the possibility of further modifications for improving the biological efficacy like the placement of a β-aryl turn at the γ-aminobutyric acid turn in the hairpin structure of PIP.

3.4 Experimental Section

3.4.1 Materials and Methods

All reagents and solvents used in this study were purchased from standard suppliers and were used without further purification. Abbreviations used here: P, *N*-methylpyrrole; I, *N*-methylimidazole; Fmoc, 9-fluorenylmethoxycarbonyl; HCTU, 1-[bis(dimethylamino)methylene]-5-chloro-1H-benzothiazolium 3-oxide hexafluorophosphate; DMSO, dimethylsulfoxide; DMF, dimethylformamide; NMP, 1-Methyl-2-pyrrolidone; DIEA, *N,N*-diisopropylethylamine; β-alanine; γ-aminobutyric acid; SAHA, suberoylanilide hydroxamic acid, HpβCD, 2-hydroxypropyl-β-cyclodextrin; Pd-C, Palladium on carbon (10% of loading) and Dp, *N,N*-dimethylaminopropylamine. 1H NMR spectra were recorded on a JEOL JNM-FX 600 model NMR spectrometer. HPLC analysis was performed with a 4.6 $\times$ 150 mm column on a JASCO PU-2089 plus pump model with UV-2075 plus HPLC UV/VIS detector and a Chemcobond 5-ODS-H 10 $\times$ 150 mm column (Chemco Scientific Co., Ltd, Osaka, Japan) was used for the purification of polyamide conjugates. Electrospray Ionization Time-of-Flight mass (ESI-TOF-MS) was recorded on BioTOF II ESI-TOF Bruker Daltonics Mass Spectrometer (Bremen, Germany). Flash column system was performed using CombiFlash Companion model (Teledyne Isco Inc., NE, USA).

3.4.2 Synthesis of Polyamides

Synthesis of SAHA-*β*-*β*-Im-Py-Py-Py-*γ*-Im-Py-Py-Py-Dp (1). Using a solid-phase peptide synthesizer, conjugate 1 was synthesized as mentioned before [10]. After SPPS, it was detached from the resin using *N*,*N*-dimethylaminopropylamine (Dp) and stirred at 45 °C for 3 h. The reaction mixture was filtered, triturated from Et_2O, to yield a yellow crude powder (21 mg). Without further purification, the precursor was dissolved in 150 μL of NMP and later another 150 μL of aqueous solution of 50% (v/v) NH_2OH were added, to convert the methylester of SAHA to the active NHOH and this allows the formation of 1. The reaction was quenched by adding 150 μL of acetic acid. The crude 1 was then purified by flash column chromatography. HPLC was checked to confirm the purity (elution with trifluoroacetic acid and a 0–100% acetonitrile linear gradient (0–40 min) at a flow rate of 1.0 mL/min under 254 nm). Yield (5 mg, 23%) (ESI-TOF-MS (positive) m/z calcd for $C_{76}H_{97}N_{25}O_{15}{}^{2+}$ $[M + 2H]^{2+}$ 799.87; found 799.89).

Synthesis of IPA(OBn) (4). Isophthalic acid (1.66 g, 10 mmol) was dissolved in a mixture of 15 mL methanol and 1 mL water and was stirred at room temperature. To the stirred solution, triethylamine (1.4 mL, 10.1 mmol) dissolved in methanol (10 mL) was added. The reaction mixture was stirred at room temperature for 12 h. The solvent was removed by vacuum pump and the residue was dissolved in 15 mL of DMF. Benzyl bromide (1.3 mL, 11 mmol) was added very slowly and the reaction mixture was stirred for another 2 h at 100 °C. The reaction was quenched by dropping the temperature to 0 °C and by adding 5% sodium bicarbonate (50 mL). The reaction mixture was then extracted with EtOAc (3 × 30 mL). The aqueous layer was carefully acidified with 10% HCL to pH 5–6 and then was extracted with EtOAc (2 × 30 mL) to remove the diester. The aqueous part was further acidified to pH 3 and again extracted with EtOAc (3 × 30 mL) to achieve 4. To the organic layer, Na_2SO_4 was added and then concentrated under reduced pressure to yield a white powder (1.18 g, 46%). TLC was checked to confirm the purity. 1H NMR (600 MHz, [d6] DMSO): δ = 8.79 (t, 1H), 8.32 (d, 1H), 8.30 (d, 1H), 8.29 (d, 1H), 7.56 (t, 2H), 7.47 (d, 2H), 7.41 (m, 1H), 5.40 (s, 2H).

Synthesis of IPA-tail (5). To a solution of 4 (300 mg, 1.17 mmol) dissolved in DMF (1 mL), PyBOP (912 mg, 1.75 mmol) was added and stirred at room temperature for 5 min. To the reaction mixture 3,3-diamine *N*-methyldipropylamine (0.282 mL, 1.75 mmol) was added and stirred at room temperature for about an hour. TLC was analyzed to check the status of the reaction and to confirm the complete disappearance of starting material. The reaction mixture was dissolved in 5% HCL and extracted with EtOAc (15 mLx2). To the aqueous layer, saturated

$NaHCO_3$ was added very slowly at ice bath till pH levels reach 12–14. Again it was extracted with EtOAc (15 mLx2) and then concentrated by rotavapor. The purity was checked using HPLC (elution with trifluoroacetic acid and a 0–100% acetonitrile linear gradient (0–20 min) at a flow rate of 1.0 mL/min under 254 nm). Yield (358 mg, 80%) (ESI-TOF-MS (positive) m/z calcd for $C_{22}H_{29}N_3O_3$ $[M + H]^+$ 384.23; found 384.20. 1H NMR (600 MHz, [d6] DMSO): $\delta = 8.66$ (s, 1H), 8.46 (s, 1H), 8.10 (d, 2H),8.06 (d, 2H), 7.47 (s, 1H), 7.37 (m, 2H), 7.28 (m, 2H), 5.31 (s, 2H), 3.49 (br, 2H) 3.39 (s, 2H), 3.12 (br, 2H) 2.89 (s, 2H), 2.74 (s, 3H), 2.04 (br s, 2H), 1.97 (br s, 2H).

Boc-*β*-*β*-Im-Py-Py-Py-*γ*-Im-Py-Py-Py-COOH (6). After the solid phase synthesis, the resin (60 mg) was added to 400 μL of 2 M NaOH and later 100 μL of 1,4-dioxane was added and were stirred at 55 °C for 3 h. The resin was filtered out and the filtrate was concentrated by vacuum pump. To the crude yellow paste-like material, Et_2O was added for precipitation with very little amount of H_2O. After washing 2–3 times with Et_2O and subsequent centrifugation, a powder form of the compound 6 (20 mg) could be obtained. ESI-TOF-MS (positive) m/z calcd for $C_{61}H_{74}N_{21}O_{14}$ $[M + H]^+$ 1324.57; found 1324.98.

Boc-*β*-*β*-Im-Py-Py-Py-*γ*-Im-Py-Py-Py-L-IPA(OBn) (7). Conjugate 6 (17 mg) was dissolved in 0.5 mL of DMF. To it, PyBOP (3 equiv.) and DIEA (3 equiv.) were added, stirred for 10 min at room temperature. Then 5 (5 equiv.) were added to the reaction mixture and stirring was continued for 2 h. After the completion of the reaction, the solvent was removed by vacuum pump and excess Et_2O were added for precipitation. The crude product 7 was washed several times with DCM and Et_2O to get a powder form. The resulting mixture was dried in a desiccator and HPLC analysis was done to confirm the purity. ESI-TOF-MS (positive) m/z calcd for $C_{83}H_{102}N_{24}O_{16}$ $[M + 2H]^{2+}$ 845.395; found 845.5.

SAHA-*β*-*β*-Im-Py-Py-Py-*γ*-Im-Py-Py-Py-L-IPA (2). Conjugate 7 (12 mg) was dissolved in methanol (3 mL). To it, Pd-C (5 mg) was added and then H_2 gas was passed into the reaction vessel with a balloon and was stirred at room temperature for 12 h. It was then filtered and concentrated using vacuum pump. In the next step, 10% DCM in TFA was used to cleave Boc group. After the Boc cleavage, it was then coupled with HOOC-SAHA (Ome) using PyBOP and DIEA. At the final step, the crude product having a methyl ester were dissolved in 150 μL of NMP and another 150 μL of aqueous 50% (v/v) NH_2OH was added and then the reaction mixture was stirred at room temperature for 12 h. The reaction was quenched with acetic acid and the product was purified by HPLC. Final purity was checked using HPLC (elution with trifluoroacetic acid and a 0–100% acetonitrile linear gradient (0–40 min) at a flow rate of 1.0 mL/min under 254 nm) ($t_R = 19.4$ min). Yield (2.7 mg, 13.5%). ESI-TOF-MS (positive) m/z calcd for $C_{86}H_{106}N_{26}O_{18}$ $[M + 2H]^{2+}$ 895.41; found 895.38.

Ac-Im-Py-Py-Py-*γ*-Im-Py-Py-Py-Dp (3). After SPPS, 3 having a free amine group at the N-terminus end was washed with 20% chloroacetic anhydride/DMF for

30 min for capping. It was then detached from the resin using *N,N*-dimethylaminopropylamine (Dp) and was stirred at 45 °C for 3 h to form 3. Final purification was done by flash column chromatography. Purity was checked by HPLC (elution with trifluoroacetic acid and a 0–100% acetonitrile linear gradient (0–40 min) at a flow rate of 1.0 mL/min under 254 nm) (t_R = 15.6 min). Yield (2.7 mg, 14%). ESI-TOF-MS (positive) m/z calcd for $C_{57}H_{69}N_{21}O_{10}$ $[M + 2H]^{2+}$ 604.80; found 604.79.

3.4.3 Solubility Analysis Using HPLC

Stock solutions of PIPs 1 and 3 with a concentration of 4 mM each were prepared in DMSO. A 0.5 μL from each polyamide were added to 500 μL of DMSO (solution A), 500 μL of PBS pH 7.00 (solution B) and 500 μL of PBS containing 50 mM of HP*β*CD (solution C) to achieve 4 μM solutions. Each solution was vortexed and then solution B and C were sonicated for 10 min. Next, B and C were allowed to equilibrate for 2 h at the room temperature and then were centrifuged for 10 min at 10000 rpm. 15 μL supernatant of each solution was analyzed using HPLC (elution with trifluoroacetic acid and a 0–50% acetonitrile linear gradient (0–20 min) at a flow rate of 1.0 mL/min under 254 nm. Solubility was calculated from the HPLC peak area of the respective polyamides and 100% solubility was considered for solution A.

3.4.4 Cell Culture and Polyamide Treatment to MEF

Mouse embryonic fibroblasts or MEF cells (C57BL/6) was purchased from American type culture collection (ATCC). About 1.5×10^5 cell/mL of MEF cells were seeded in a 30 mm dish culture plate containing DMEM medium. After overnight incubation for attachment, the medium was replaced with fresh DMEM (2 mL) containing polyamide conjugates in 100 nM concentrations. In the CD study, first 1 was dissolved in 1:1 mixture of 2X PBS and in either 5 or 50 mM of CD. Then sonication was done for 10 min followed by equilibration for about an hour prior to the treatment in MEF. DMSO (0.1%) treated MEF was used for the control experiment. All the plates were incubated in a 5% CO_2 humidified atmosphere at 37 °C. MEF treated with effectors were harvested at 24 h. Mouse embryonic stem (ES) cell lines R1 is used as a positive control in this study and is a kind gift from Prof. Nakatsuji lab, iCeMS, Kyoto University.

3.4.5 *Quantification of Expression of Marker Genes in MEF*

Total RNA was extracted from ES cells and 1, 2 and DMSO-treated MEF using an RNeasy Mini Kit (Qiagen) and cDNA was synthesized with ReverTra Ace qPCR RT kit (Toyobo, Japan). SYBR green real-time RT-PCR amplifications were carried out in triplicate with the protocol and conditions mentioned in THUNDERBIRD SYBR qPCR Mix (TOYOBO, Japan) on an ABI 7300 Real-Time Detection System (Applied Biosystems, USA) and were analyzed using a 7300 System SDS Software v1.3.0 (Applied Biosystems, USA) as mentioned before. After normalization with the housekeeping gene GAPDH, using the comparative cycle threshold (CT) method, the relative expression level of each gene was analyzed by considering the gene expression in DMSO treated cells as onefold. As shown in the previous reports [10, 20], primer pairs for the endogenous genes such as Oct-3/4, Nanog, Sox2, Klf4, Dppa4, Cdh1, and Rex1 were used. Statistical analysis was done using GraphPad Prism software and the results are shown as mean values ± standard deviation (SD). Statistical significance was determined by *t*-test and *p*-values were obtained with two-tailed method. The *p*-values less than 0.05 reflect the significance of the difference and the values more than that is assigned as non-significant (NS).

References

1. Hou P, Li Y, Zhang X, Liu C, Guan J, Li H, Zhao T, Ye J, Yang W, Liu K, Ge J, Xu J, Zhang Q, Zhao Y, Deng H (2013) Pluripotent stem cells induced from mouse somatic cells by small-molecule compounds. Science 341:651–654. https://doi.org/10.1126/science.1239278
2. (a) Wu YL, Pandian GN, Ding YP, Zhang W, Tanaka Y, Sugiyama H (2013) Clinical grade iPS cells: need for versatile small molecules and optimal cell sources. Chem Biol 20:1311–1322. https://doi.org/10.1016/j.chembiol.2013.09.016; (b) Pandian GN, Sugiyama H (2012) Programmable genetic switches to control transcriptional machinery of pluripotency. Biotechnol J 7:798–809. https://doi.org/10.1002/biot.201100361
3. Ohtsuki A, Kimura MT, Minoshima M, Suzuki T, Ikeda M, Bando T, Nagase H, Shinohara K, Sugiyama H (2009) Synthesis and properties of PI polyamide—SAHA conjugate. Tetrahedron Lett 50:7288–7292. https://doi.org/10.1016/j.tetlet.2009.10.034
4. Wähnert U, Zimmer O, Luck G, Pitra O (1975) (dA-dT) Dependent inactivation of the DNA template properties by interaction with netropsin and distamycin A. Nucleic Acids Res 2: 391–404
5. Kopka ML, Yoon C, Goodsell D, Pjura P, Dickerson RE (1985) Binding of an antitumor drug to DNA, Netropsin and C-G-C-G-A-A-T-T-BrC-G-C-G. Mol Biol 183:553–563
6. Dervan PB (2001) Molecular recognition of DNA by small molecules. Bioorg Med Chem 9:2215–2235
7. Han L, Pandian GN, Junetha S, Sato S, Anandhakumar C, Taniguchi J, Saha A, Bando T, Nagase H, Sugiyama H (2013) A synthetic small molecule for targeted transcriptional activation of germ cell genes in a human somatic cell. Angew Chem Int Ed 52:13410–13413. https://doi.org/10.1002/anie.201306766

8. (a) Pandian GN, Taniguchi J, Junetha S, Sato S, Han L, Saha A, AnandhaKumar C, Bando T, Nagase H, Vaijayanthi T, Taylor RD, Sugiyama H (2014) Distinct DNA-based epigenetic switches trigger transcriptional activation of silent genes in human dermal fibroblasts. Sci Rep 4:3843. https://doi.org/10.1038/srep03843; (b) Pandian GN, Sugiyama H (2012) Strategies to modulate heritable epigenetic defects in cellular machinery: lessons from nature. Pharmaceuticals 6:1–24. https://doi.org/10.3390/ph6010001
9. Pandian GN, Shinohara K, Ohtsuki A, Nakano Y, Masafumi M, Bando T, Nagase H, Yamada Y, Watanabe A, Terada N, Sato S, Morinaga H, Sugiyama H (2011) Synthetic small molecules for epigenetic activation of pluripotency genes in mouse embryonic fibroblasts. Chem BioChem 12:2822–2828. https://doi.org/10.1002/cbic.201100597
10. Pandian GN, Nakano Y, Sato S, Morinaga H, Bando T, Nagase H, Sugiyama H (2012) A synthetic small molecule for rapid induction of multiple pluripotency genes in mouse embryonic fibroblasts. Sci Rep 2:544. https://doi.org/10.1038/srep00544
11. Saha A, Pandian GN, Sato S, Taniguchi J, Hashiya K, Bando T, Sugiyama H (2013) Synthesis and biological evaluation of a targeted DNA-binding transcriptional activator with HDAC8 inhibitory activity. Bioorg Med Chem 21:4201–4209. https://doi.org/10.1016/j.bmc.2013.05.002
12. Jacobs CS, Dervan PB (2009) Modifications at the C-terminus to improve pyrrole-imidazole polyamide activity in cell culture. J Med Chem 52:7380–7388. https://doi.org/10.1021/jm900256f
13. Hargrove AE, Raskatov JA, Meier JL, Montgomery DC, Dervan PB (2012) Characterization and solubilization of pyrrole-imidazole polyamide aggregates. J Med Chem 55:5425–5432. https://doi.org/10.1021/jm300380a
14. (a) Belitsky JM, Leslie SJ, Arora PS, Beerman TA, Dervan PB (2002) Cellular uptake of N-methylpyrrole/N-methylimidazole polyamide-dye conjugates. Bioorg Med Chem 10:3313–3318; (b) Vaijayanthi T, Bando T, Hashiya K, Pandian GN, Sugiyama H (2013) Design of a new fluorescent probe: pyrrole/imidazole hairpin polyamides with pyrene conjugation at their γ-turn. Bioorg Med Chem 21:852–855. https://doi.org/10.1016/j.bmc.2012.12.018
15. Nickols NG, Jacobs CS, Farkas ME, Dervan PB (2007) Improved nuclear localization of DNA-binding polyamides. Nucleic Acids Res 35:363–370. https://doi.org/10.1093/nar/gkl1042
16. Adlington RM, Baldwin JE, Becker GW, Chen B, Cheng L, Cooper SL, Hermann RB, Howe TJ, McCoull W, McNulty AM, Neubauer BL, Pritchard GJ (2001) Design, synthesis, and proposed active site binding analysis of monocyclic 2-azetidinone inhibitors of prostate specific antigen. J Med Chem 44:1491–1508
17. Chenoweth DM, Harki DA, Dervan PB (2009) Solution-phase synthesis of pyrrole-imidazole polyamides. J Am Chem Soc 131:7175–7181. https://doi.org/10.1021/ja901307m
18. Nishijima S, Shinohara K, Bando T, Minoshima M, Kashiwazaki G, Sugiyama H (2010) Cell permeability of Py-Im-polyamide-fluorescein conjugates: influence of molecular size and Py/Im content. Bioorg Med Chem 18:978–983. https://doi.org/10.1016/j.bmc.2009.07.018
19. Meier JL, Montgomery DC, Dervan PB (2012) Enhancing the cellular uptake of Py-Im polyamides through next-generation aryl turns. Nucleic Acids Res 40:2345–2356. https://doi.org/10.1093/nar/gkr970
20. Takahashi K, Yamanaka S (2006) Induction of pluripotent stem cells from mouse embryonic and adult fibroblast cultures by defined factors. Cell 126:663–676. https://doi.org/10.1016/j.cell.2006.07.024

Chapter 4
Development of a Novel Photochemical Detection Technique for the Analysis of Polyamide-Binding Sites

Abstract We demonstrated sequence-specific electron injection using pyrene-conjugated pyrrole–imidazole polyamide (PIP) on 5-bromouracil (^{Br}U)-substituted DNA. ^{Br}U is a good electron acceptor and is able to trap an electron from pyrene-conjugated polyamides after irradiation at 365 nm. This results in strand cleavage of the DNA by generating uracil-5-yl radical from ^{Br}U after eliminating the bromide anion. Using this exciting photochemical tool, we have analyzed the binding affinity, specificity, and binding orientation of four pyrene PIPs (1–4) using two long ^{Br}U-substituted DNAs (298 and 381 bp). The electron injection sites of 1–2 confirm very low sequence specificity, whereas 3, 4 exhibited high sequence specificity; moreover, electron injection by 2 also confirms its preference for reverse binding site. The binding affinity was further validated by SPR. Thus, these results suggest that it can be a useful tool for detecting binding sites of small molecules.

Keywords Pyrene polyamide · Strand cleavage · Intermolecular electron transfer ^{Br}U-substituted DNA · Binding affinity

4.1 Introduction

Replacement of thymine by isosteric 5-bromouracil (^{Br}U) in DNA is attractive since it does not affect the functionality of the resulting DNA due to similar sizes of Br (1.95 Å) and methyl (2 Å) [1]. As a result of this modification, it greatly increases its photosensitivity with respect to protein–nucleic acid crosslinks, [1, 2] single- and double-strand breaks, and the creation of alkali-labile sites [3–7]. The chemistry of strand cleavage after irradiation at 302 nm of ^{Br}U-labeled DNA has been investigated based on the photodegraded products of a model hexamer d(GCA^{Br}UGC)$_2$ [8]. The abstraction of hydrogen by the uracil-5-yl radical (U$^{•}$), a powerful hydrogen abstractor, generated from the ^{Br}U anion radical by eliminating the bromide ion, is the main photochemical event in such photoinduced strand cleavage. The isolation of two kinds of photodegraded products has pointed out two possible ways of hydrogen abstraction from the deoxyribose moiety at the 5′ position [9].

A. Saha, *Molecular Recognition of DNA Double Helix*, Springer Theses,
https://doi.org/10.1007/978-981-10-8746-2_4

The abstraction of hydrogen from the C1′ position generates 2-deoxyribonolactone with the release of A, and abstraction from the C2′α position generates an erythrose-containing site [9]. The detailed mechanism of photodamage in ^{Br}U-labeled DNA has recently been clarified further with the identification of the hotspot sequences 5′–G/C[A]$_{n=1,2,3}$ ^{Br}U^{Br}U–3′ and its reverse 3′ → 5′ sequence [10–12]. Initially, the A residue adjacent to ^{Br}U was considered to be the electron donor; however, subsequent experiments confirmed that the G residue, which has the lowest oxidation potential among the four bases, is the electron donor [8, 11]. Moreover, the A bridges help to prevent rapid back electron transfer [11]. In contrast to our observations, direct strand cleavage has also been reported after aerobic and anaerobic photolysis [13, 14]. The proposed mechanism for direct strand breakage also involves hydrogen abstraction by U• from the C1′ and C2′ positions at 5′, albeit via different mechanistic pathways. Recently, successful quenching of highly reactive U• in the hotspot sequences was shown by supplementation with an excess amount of the hydrogen donor tetrahydrofuran (THF), to stop intra-strand hydrogen abstraction [12]. The elucidation of the mechanistic details of the photochemistry of ^{Br}U-labeled DNA has attracted much attention to its biological and technological applications.

Pyrrole–Imidazole Polyamides (PIPs) are synthetic small molecules that can bind to the minor groove in a sequence-specific manner. This molecule gained much interest in biology since it has been shown to regulate gene expression [15]. Footprinting and affinity cleavage techniques have guided the development of pairing rule of PIPs and it is likely to be the side by side stacked ring pairing, such as an Im opposite to Py (Im/Py), can recognize a G/C base pair from the C/G base pair, whereas a Py located opposite to Py (Py/Py) recognizes either an A/T or T/A base pair [15–21]. It is common to see a polyamide binding preferentially to a high-affinity binding site in the presence of several potential binding sites, although the underlying mechanism remains unclear. Therefore, it is crucial to screen a library of potential binding sites to identify real highest affinity binding sites. The application of PIPs in biology will require knowledge of its sequence specificity and has necessitated a more advanced screening technique that can be routinely used in higher throughput. Although affinity cleavage methods explored affinity to a great extent, they are partially sensitive and the synthesis challenges regarding polyamide-bindingthe attachment of EDTA to polyamide render them challenging methods for the determination of binding sites. Because of easier pyrene conjugation and ^{Br}U-substituted DNA being more sensitive than native DNA, the combined output of the current approach is more facile and sensitive than previous approach.

4.2 Results

4.2.1 Analysis of Photoreacted Sample Using PAGE

To test this hypothesis, we prepared two long, fully ^{Br}U-substituted DNAs fragments (381 bp: DNA1 and 298 bp: DNA2) using PCR and designed four pyrene-conjugated polyamides (**1**–**4**, Fig. 4.1) to check their binding affinity and specificity by electron injection. In early studies, it has been well characterized that pyrene serves as an electron donor in DNA mediated excess electron transfer [22–26]. Thus, after irradiation at 365 nm for a minimum of 5 s, each pyrene PIP injected an electron to ^{Br}U residues at their binding sequences. Using high-resolution gel electrophoresis, the site of electron injection was identified from the strand break caused by the sequence-selective generation of U$^{\bullet}$ radical. A photoreaction scheme of the photoinduced electron injection in a ^{Br}U-substituted DNA is shown in Fig. 4.2.

The two DNA fragments (DNA1 and DNA2) were incubated with pyrene polyamide in sodium cacodylate buffer (pH 7.00) and isopropanol and were irradiated with 365 nm LED at 0 °C for 0–30 s. Isopropanol acts as a hydrogen donor; thus, when supplied in excess amount (500 mM), it quenched the U$^{\bullet}$ to uracil (U). The photoreacted sample was then digested with uracil-DNA glycosylase (UDG), which converted the U residue to a heat-labile abasic site. Under heating conditions (95 °C for 10 min), the abasic site was easily cleaved and the resultant strand break was then analyzed by 6% denaturing polyacrylamide gel electrophoresis (PAGE), Fig. 4.2. The results of PAGE are shown in Figs. 4.3a and 4.4a.

4.2.2 Data Analysis of Polyamide-Binding Sites from Photoreaction

Polyamide 1 (pyrene-ββPyImPyIm-γ-PyPyPyPyβDp) can recognize 5′–WCWCWW–3′ and 5′–WWGWGW–3′ (W = A/T) sequences. The analysis of DNA1 from the top and bottom strand separately revealed the presence of strand cleavage in three match sequences (site 1, 3, 5) via electron injection (shown in red in Fig. 4.3b). Moreover, two 1 bp mismatch sites (site 2 and 4) were detected by strand cleavage. The amount of strand cleavage at the mismatch sites was almost identical to that detected at the match sites. In contrast, in DNA2 (Fig. 4.4a) **1** exhibited four match sequences, and consistent strand cleavage was observed in all match sequences (site 1–4). In addition, there were two strong strand cleavage events at sites 5 and 6, which were considered as 1 bp mismatch sites. Strand cleavage at the mismatch sites of **1** was consistently strong, as revealed by both DNA fragments, which clearly indicates that **1** has low sequence specificity. Moreover, it has been noticed that the amount of strand cleavage was not identical at all match sites. This might result from sequence preference among many potential binding sites, although the exact reason is unknown.

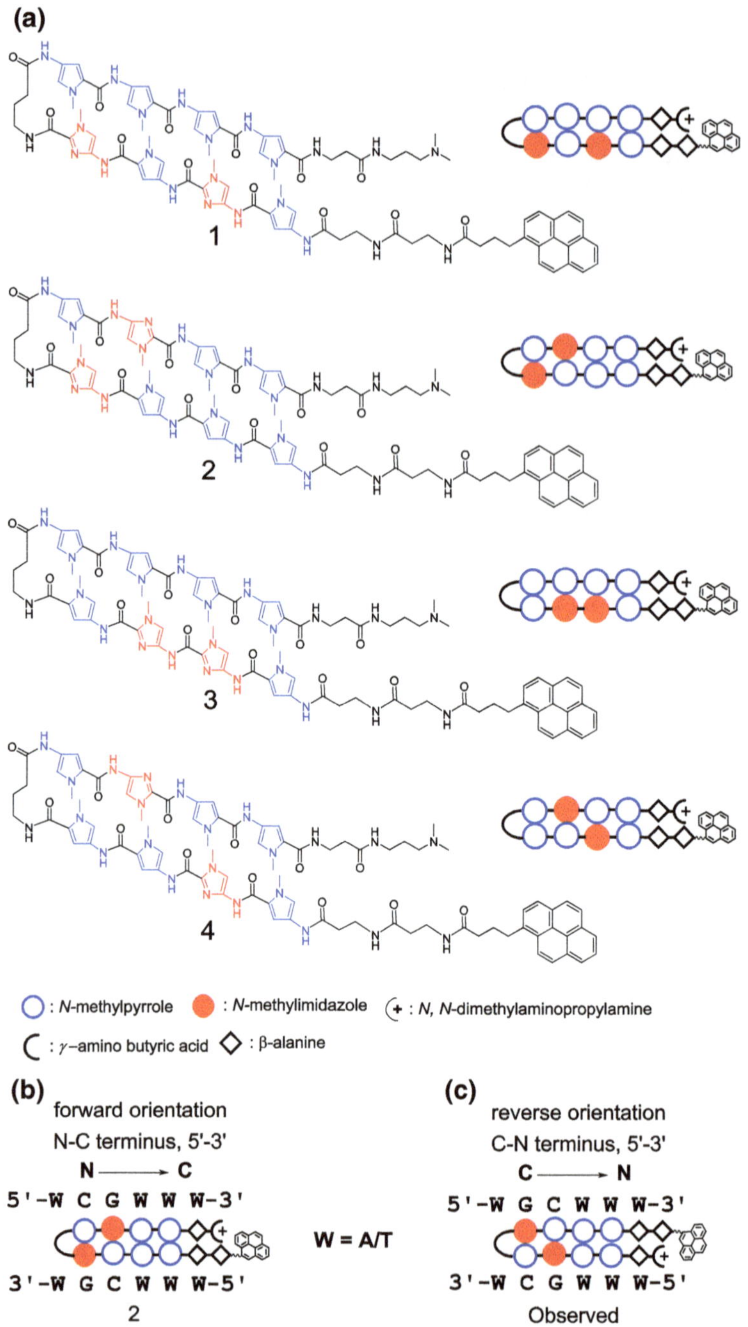

Fig. 4.1 **a** Chemical structures of pyrene polyamides **1–4** in the forward orientation. **b** Forward binding orientation in which the N–C terminus of polyamides aligns with 5′ → 3′ DNA. **c** Reverse binding orientation in which the C–N terminus of polyamides aligns with 5′ → 3′ DNA. A reverse orientation for polyamide 2 is shown, which was detected in this photoaffinity cleavage method

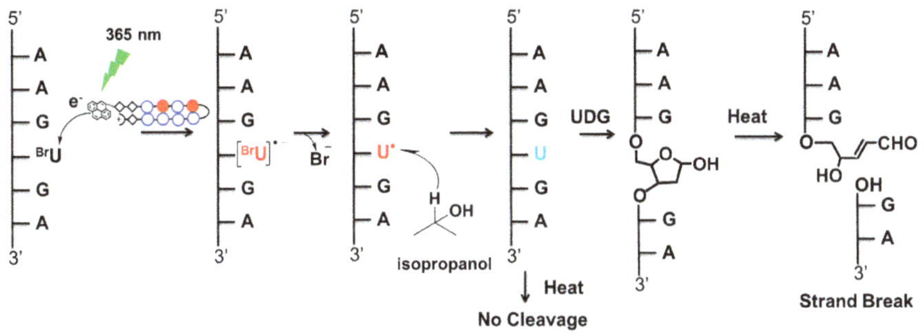

Fig. 4.2 Reaction scheme used to detect strand cleavage via electron injection using pyrene-conjugated pyrrole–imidazole polyamide in ^{Br}U-substituted DNA. A pyrene polyamide that can bind to the minor groove can be targeted to bind at specific sites and under photoirradiation, the pyrene moiety of the polyamide would be excited to transfer an electron to ^{Br}U residues at its binding site. ^{Br}U is an efficient electron acceptor that generates the highly reactive uracil-5-yl radical (shown in red) by capturing an electron. In the presence of excess hydrogen atom donor, isopropanol, the radical species can be quenched to uracil (shown in blue). Treatment with the uracil-DNA glycosylase (UDG) enzyme can convert uracil to heat-labile Ap site selectively, which under heating conditions (at 95 °C) causes strand breakage

Polyamide 2 (pyrene-ββPyPyPyIm-γ-PyImPyPyβDp) can recognize 5′–WCG WWW–3′ and 5′–WWWCGW–3′ sequences. However, in the DNA1 fragment, it did not have any match sequence. But, the photoreaction gives several strong strand cleavage events at sites 6, 7, 9, 10, 11, which are considered as 1 bp mismatch sites, Fig. 4.3a, c. Importantly, a strand cleavage at site 8 shifted our attention toward the reverse orientation of Polyamide 2. Surprisingly, the sequence of cleavage is a 1 bp mismatch reverse binding site of **2** (marked in green). Generally, polyamides preferentially align with the N terminus of the antiparallel DNA strand, with its 5′ end in the N → C, 5′ → 3′ orientation, which is known as the forward orientation. In some cases, polyamides can also recognize the reverse orientation, for example, C → N, 5′ → 3′, as shown in Fig. 4.1b, c [27]. Strand cleavage in the DNA2 fragment, Fig. 4.4a, was more interesting. Similarly, a reverse orientation binding was observed at site 10, which is a full match reverse binding site of **2**, Fig. 4.4c. The amount of strand cleavage (band intensity) at site 10 was almost identical to that observed at match sites. This implies that **2** can bind in the reverse orientation efficiently. Since polyamide 2 have match sites in DNA2, thus it gave cleavage at match sites, such as sites 7 and 9. In addition, some 1 bp mismatch sites were also observed such as sites 8 and 11. It should be noted that photoirradiated DNA1-2 fragments incubated with **2** consumed almost nearly 90% of the total amount of unreacted DNA in a condition that was similar to that used for other polyamides (Figs. 4.3a and 4.4a). This suggests nonspecific binding of polyamide 2.

Polyamide 3 (pyrene-ββPyImImPy-γ-PyPyPyPyβDp), can target 5′–WWG GWW–3′ and 5′–WWCCWW–3′ sequences. The location of these sequences in the DNA1 was completely different from the cleavage sites of **1** and **2**. Thus, strand

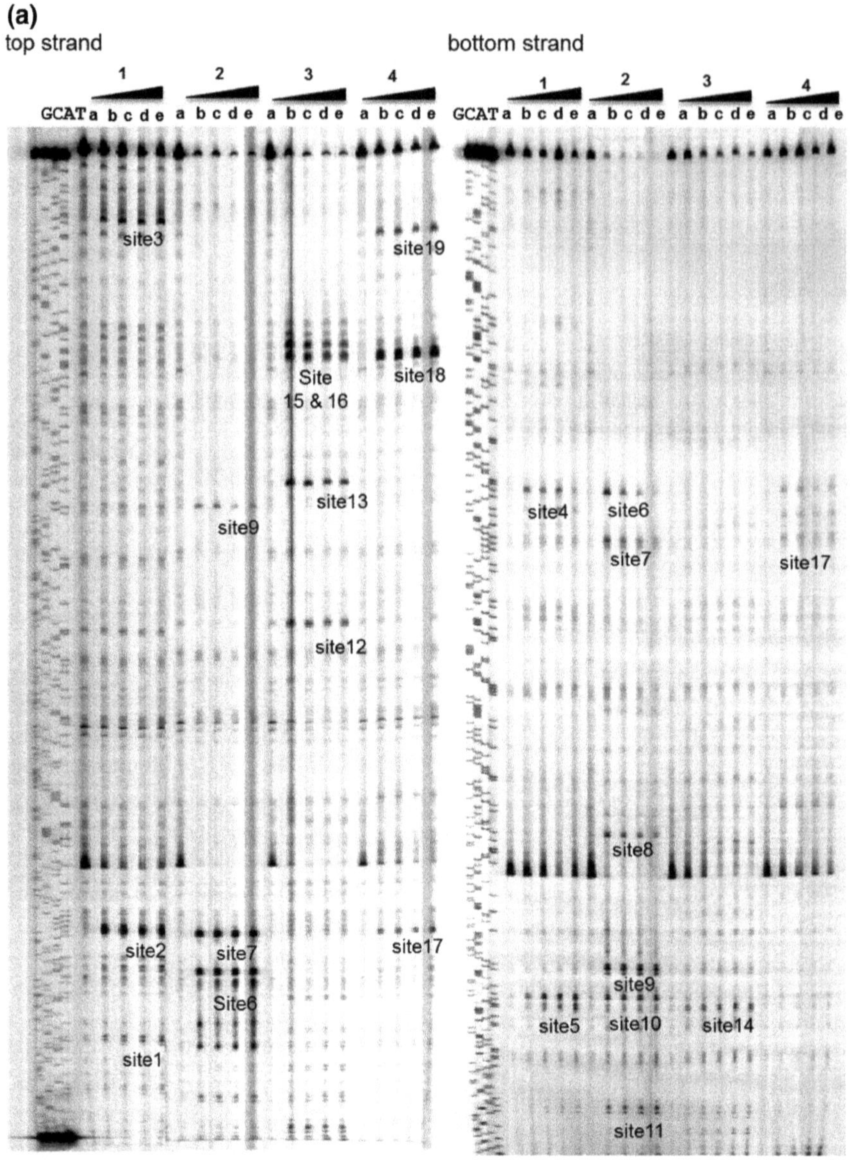

Fig. 4.3 **a** Slab gel sequencing analysis of a 6% denaturing (7 M urea) polyacrylamide gel for the DNA1 (381 bp) after photoreaction using polyamides **1**–**4**. Lanes a–e, photoirradiation period of 0, 5, 10, 15, and 30 s, respectively, at 365 nm UV using 10 nm DNA and 100 nm polyamide. Top- and bottom-strand analyses are shown on the left and right, respectively. **b** Mapping of the electron injection site from the gel as shown in Fig. 4.3a for DNA1 in both the top and bottom strands. **c** Electron injection sites for polyamides **1**, **2**, **3**, and **4** at sites 1–5, 6–11, 12–16, and 17–19, respectively. The binding sites are shown in different colors: red indicates a match site, blue indicates a 1 bp mismatch site, and green indicates a reverse binding site; the boxes at mismatch sites indicate the mismatched base pair. Arrows indicate the strand break site, with intensity

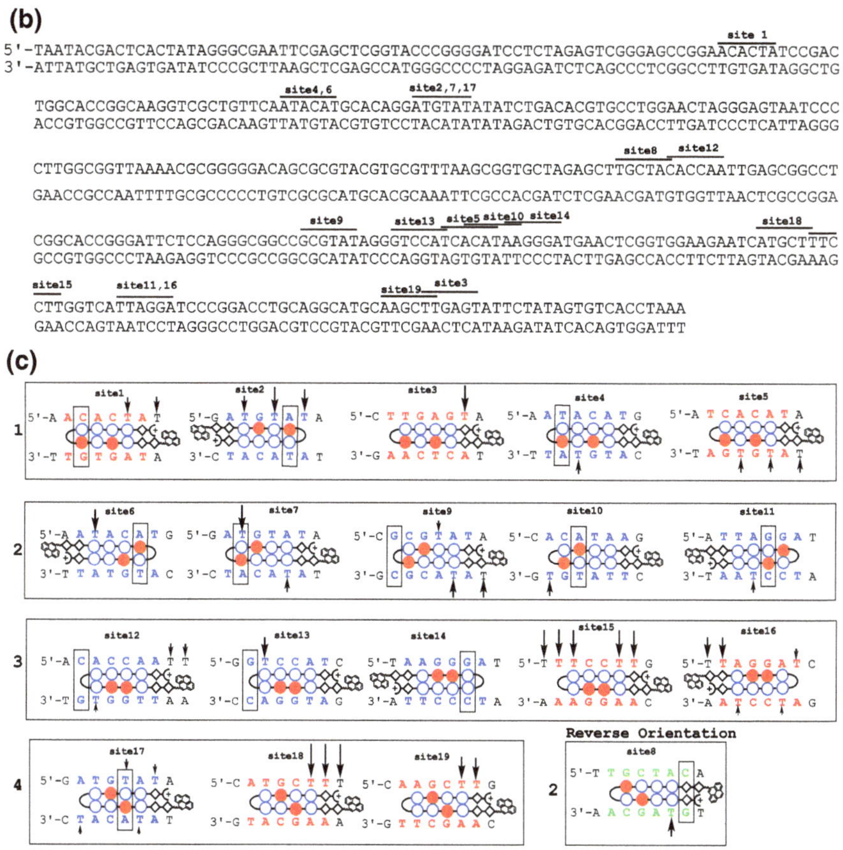

Fig. 4.3 (continued)

cleavage at the target site of **3** was deemed as crucial to support electron injection based on sequence specificity. Interestingly, strong strand cleavage at the target sites of **3**, such as sites 15 and 16, Fig. 4.3a, was detected, which strongly implies electron injection through sequence selectivity of **3**. As polyamides targeting 6 bp DNA sequences can also bind to mismatch sequences, cleavage at 1 bp mismatch sites, such as sites 12–14, by **3** was normal. However, the amount of cleavage at these sites was lower than that detected at match sites, which also indicates that **3** can differentiate between match and mismatch sites under a similar condition. In contrast, in the DNA2 fragment there was no match sequence; however, three extremely weak strand cleavage events, which were considered as 1 bp mismatch sequences (sites 12–14, Fig. 4.4a, c), were observed. This observation was consistent in both DNA fragments.

Polyamide 4 (pyrene-ββPyImPyPy-γ-PyImPyPyβDp) In a similar way, we designed polyamide **4**, to target 5′–WWGCWW–3′ sequences because of their unique places in both DNA fragments. Strand cleavage events at sites 18 and 19 confirmed its match sequences in the DNA1 fragment as shown in Figs. 4.3 and 4.4. Moreover, a weak strand cleavage observed at site 17 confirmed its 1 bp mismatch sequence. In the DNA2, site 15 was the only match sequence for **4** in the entire DNA fragment, and strong strand cleavage at that sequence further confirmed the sequence selectivity. In addition, two 1 bp mismatch sequences, sites 16 and 17, were also cleaved after the photoirradiation. It should be noted that, although these types of polyamides can easily bind to 1 bp mismatch sequences, the number of mismatches and the amount of cleavage in the case of **4** were lower. This implies its higher sequence specificity among the four designed polyamides. Again it is to be noted that whether pyrene has any influence on polyamide-binding specificity or not is unknown. However, pyrene is known to bind DNA relatively, nonspecifically by intercalating between the base pairs [28]; the absence of such cleavages suggest that pyrene conjugation with polyamide might not influence the specificity. Because in the present study we admitted that binding affinity, binding orientation, and specificity can be estimated by strand cleavage via electron injection, we performed SPR on biotinylated DNA [29] containing those particular sequences of match, mismatch, and reverse binding sites for further validation of the results.

4.2.3 SPR Analysis

First, we examined the binding affinity of polyamide **1** at match and 1 bp mismatch sites. The results are shown in Table 4.1, Fig. 4.5. The K_D values for match and mismatch sites were 5.96×10^{-8} and 1.05×10^{-7} (*M*), respectively. The specificity between these two values was just 1.76. These values indicate almost a similar binding affinity in match and mismatch sites that were observed from strand cleavage in both DNA fragments. As photoirradiation of DNA incubated with **2** pointed out the possibility of binding in the reverse orientation, binding affinity in forward and reverse orientations was further examined by SPR. The K_D value in the forward and reverse orientations of **2** were 5.13×10^{-8} and 1.51×10^{-7} (*M*), respectively, and the specificity was just 2.94. This result suggests that **2** can bind in the reverse orientation at site 10 in the DNA2 fragment with almost the same affinity as that detected in the forward orientation; see Fig. 4.4a, b. The ~3.5 times higher k_a value observed in the reverse site compared with the forward site also indicates that **2** can access the reverse sequence more efficiently. We also checked the reverse binding affinity of **1**, as it contains several reverse binding sites in both DNA fragments. However, those reverse sequences overlapped with forward sequences, for example, site 1 in Fig. 4.3b. Polyamide **1** exhibited a K_D value of 5.57×10^{-7} (*M*) in the reverse orientation, and its specificity with respect to the forward orientation was 9.34. This value indicates a significant low binding affinity for **1** in the reverse orientation. Moreover, the k_a value for the forward orientation was ~3.5 times higher than that for the reverse orientation, which is completely opposite to what was observed for **2**.

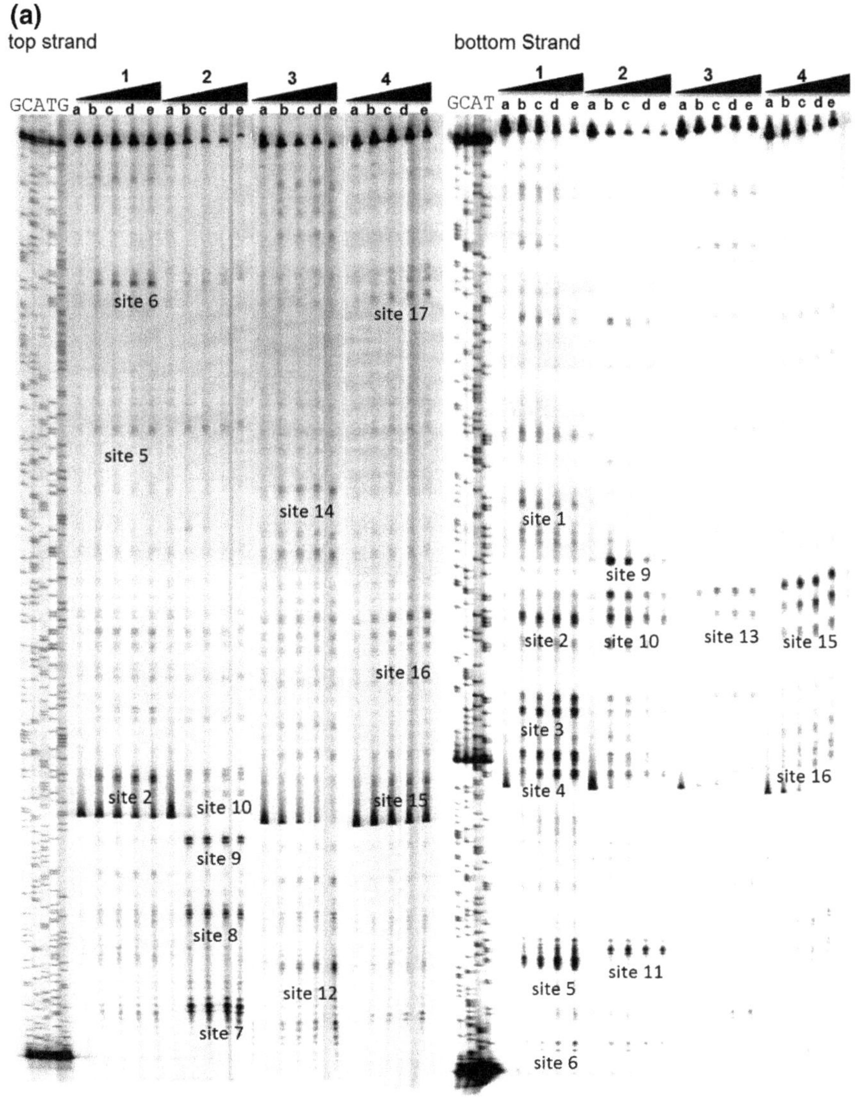

Fig. 4.4 **a** Slab gel sequencing analysis of a 6% denaturing (7 M urea) polyacrylamide gel for the DNA2 (298 bp) after photoreaction using polyamides **1**–**4**. Lanes a–e, photoirradiation period of 0, 5, 10, 15, and 30 s, respectively, at 365 nm UV using 10 nm DNA and 100 nm polyamide. Top- and bottom-strand analyses are shown on the left and right, respectively. **b** Mapping of the electron injection site from the gel as shown in Fig. 4.4a for the DNA1 in both the top and bottom strands. **c** Electron injection sites for polyamides **1**, **2**, **3**, and **4** at sites 1–6, 7–11, 12–14, and 15–17, respectively. The binding sites are shown in different colors: red indicates a match site, blue indicates a 1 bp mismatch site, and green indicates a reverse binding site; the boxes at mismatch sites indicate the mismatched base pair. Arrows indicate the strand break site, with intensity

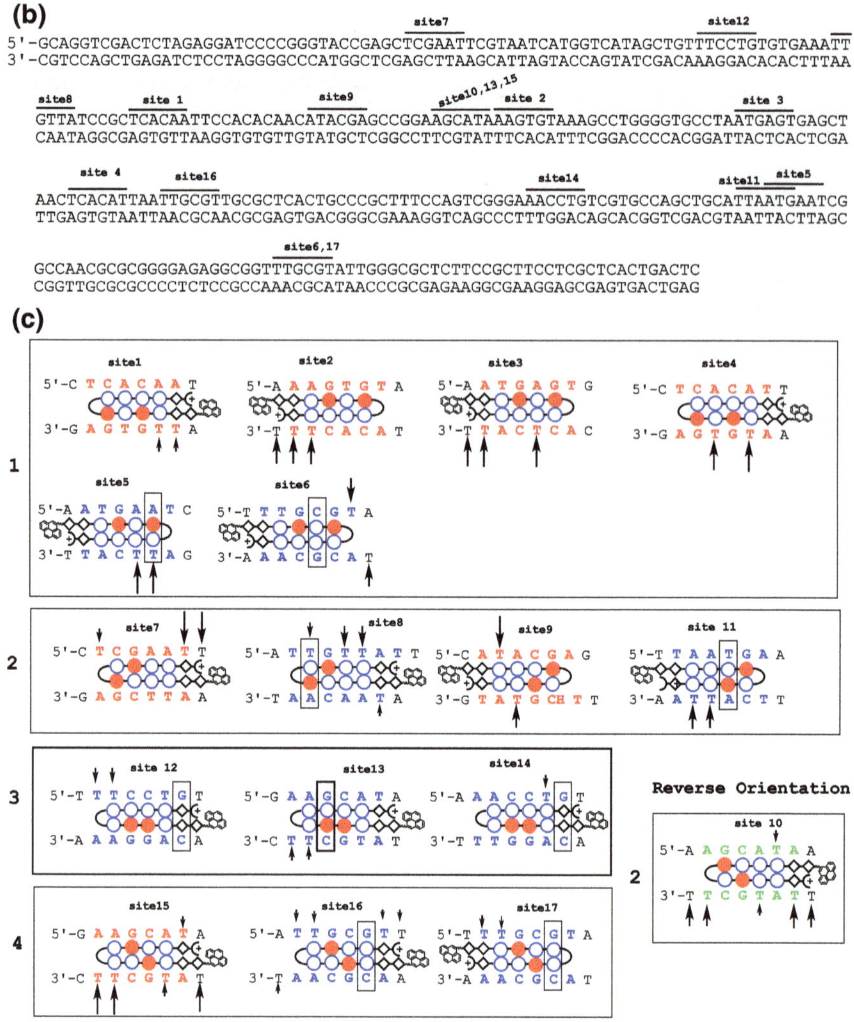
(b)
site7
site12
5'-GCAGGTCGACTCTAGAGGATCCCCGGGTACCGAGCTCGAATTCGTAATCATGGTCATAGCTGTTTCCTGTGTGAAATT
3'-CGTCCAGCTGAGATCTCCTAGGGGCCCATGGCTCGAGCTTAAGCATTAGTACCAGTATCGACAAAGGACACACTTTAA
site8
site 1
site9
site10,13,15
site 2
site 3
GTTATCCGCTCACAATTCCACACAACATACGAGCCGGAAGCATAAAGTGTAAAGCCTGGGGTGCCTAATGAGTGAGCT
CAATAGGCGAGTGTTAAGGTGTGTTGTATGCTCGGCCTTCGTATTTCACATTTCGGACCCCACGGATTACTCACTCGA
site 4
site16
site14
site11
site5
AACTCACATTAATTGCGTTGCGCTCACTGCCCGCTTTCCAGTCGGGAAACCTGTCGTGCCAGCTGCATTAATGAATCG
TTGAGTGTAATTAACGCAACGCGAGTGACGGGCGAAAGGTCAGCCCTTTGGACAGCACGGTCGACGTAATTACTTAGC
site6,17
GCCAACGCGCGGGGAGAGGCGGTTTGCGTATTGGGCGCTCTTCCGCTTCCTCGCTCACTGACTC
CGGTTGCGCGCCCCTCTCCGCCAAACGCATAACCCGCGAGAAGGCGAAGGAGCGAGTGACTGAG
(c)
1
site1
site2
site3
site4
site5
site6
2
site7
site8
site9
site 11
3
site 12
site13
site14
4
site15
site16
site17
Reverse Orientation
2
site 10

Fig. 4.4 (continued)

Table 4.1 SPR analysis validated the binding affinity

Pyrene polyamide	k_a $(M^1 s^{-1})$	k_d (s^{-1})	K_D (M)	χ^2	Specificity
PIP 1- Match 5′-Biotin-GCGCTTGAGTCGCGT T  CGCGAACTCAGCGCT T	1.96×10^4	1.17×10^{-3}	5.96×10^{-8}	0.27	–
PIP 1-Mismatch 5′-Biotin-GCGCATGTATCGCGT T  CGCGTACATAGCGCT T	2.35×10^3	2.46×10^{-4}	1.05×10^{-7}	0.62	1.76
PIP 1-Reverse 5′-Biotin-GCGCTTCACTCGCGT T  CGCGAAGTGAGCGCT T	5.55×10^3	3.09×10^{-3}	5.57×10^{-7}	1.75	9.34
PIP 2-Match 5′-Biotin-GCGCATACGACGCGT T  CGCGTATGCTGCGCT T	6.23×10^3	3.19×10^{-4}	5.13×10^{-8}	0.47	–
PIP 2-Reverse 5′-Biotin-GCGCTATGCTCGCGT T 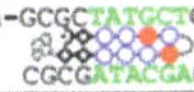 CGCGATACGACGGCT T	2.19×10^4	3.31×10^{-3}	1.51×10^{-7}	1.24	2.94

4.3 Conclusion

In this chapter, we have demonstrated that the photochemistry of ^{Br}U-substituted DNA can be used to detect the binding site of PIPs. Using this technique, we have successfully analyzed the binding sites of four pyrene-conjugated PIPs in two long ^{Br}U-substituted DNA fragments (381 and 298 bp) by photoinduced electron injection and estimated their binding affinity, specificity, and orientation preferences. Using high-resolution denaturing gel electrophoresis, the site of electron injection was analyzed. PAGE analysis revealed a unique pattern of electron injection from polyamides **1**–**4** in both DNA fragments. Our results suggest that **1** and **2** inject electron with low sequence specificity and **3** and **4** inject electron with high sequence specificity. Moreover, using this technique we can also detect reverse orientation binding sites, as observed for **2**. As it is crucial to scan libraries of many potential binding sites to identify binding sites with the highest affinity, this platform would be a useful screening tool in the design and development of PIPs.

4.4 Methods

4.4.1 Preparation of TexasRed End-Labeled ^{Br}U-Substituted DNA

The DNA fragments were amplified by PCR using d^{Br}UTP instead of dTTP using respective plasmids. The DNA1 was amplified by using forward 5′-TAATACG ACTCACTATAGG-3′ and reverse 5′-ATTTAGGTGACACTATAGAATAC-3′ primers from pGEM3Z-601, and DNA2 was amplified by using forward 5′-GCAG GTCGACTCTAGAGGAT-3′and reverse 5′-GAGTCAGTGAGCGAGGAAG-3′

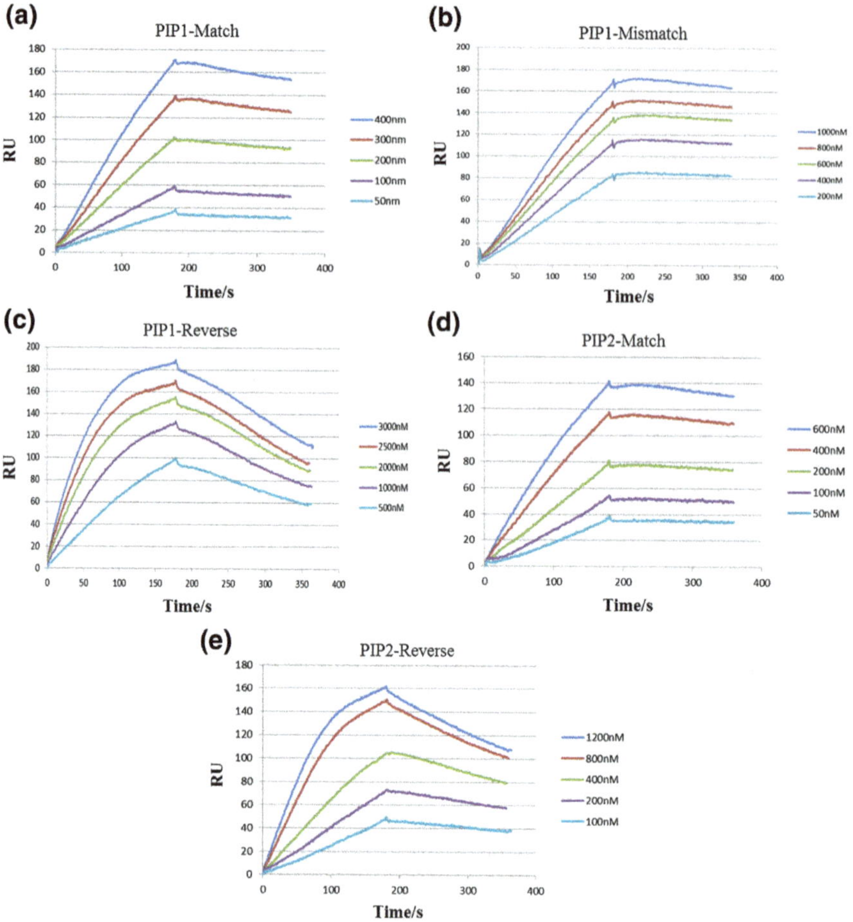

Fig. 4.5 SPR analyses of pyrene polyamides. **a** Polyamide 1 at match site **b** polyamide 1 at 1 bp mismatch site **c** polyamide at reverse binding orientation **d** polyamide 2 at match site **e** polyamide 2 at reverse binding orientation. X-axis indicates running time and Y-axis is the corresponding RU

primers from pUC18. Primers shown above, either forward or reverse having TexasRed at the 5′-end, were used for making either top strand or bottom strand labeled ^{Br}U-DNA. After PCR amplification, DNA was purified using the GenElute PCR Clean-Up Kit (Sigma-Aldrich, St. Louis, MO, USA) and quantified by Nanodrop.

4.4.2 Synthesis of Polyamides

Polyamides, **1–4**, were synthesized using a β-ala-Wang resin under solid phase peptide synthesizer PSSM-8 with Fmoc chemistry for repetitive coupling. Once the machine-assisted synthesis was finished, polyamides were detached from the resin by using Dp at 55 °C for 3 h. It was then freshly washed with cold Et_2O and without further purification it was coupled with a pyrene butyric acid *N*-hydroxysuccinimide ester to achieve the final products. Within 1 h, the reaction was finished and final purification was done by HPLC. The product was confirmed by TOF-MS and TOF Mass.

Polyamide **1** ESI-TOF Mass m/z calcd. $C_{84}H_{98}N_{24}O_{13}$ $[M + 2H]^{2+}$ 825.39; found 825.26

Polyamide **2** ESI-TOF Mass m/z calcd. $C_{84}H_{98}N_{24}O_{13}$ $[M + 2H]^{2+}$ 825.39; found 825.25

^{1}HNMR for polyamide **1** and **2** was reported in the reference 18.

Polyamide **3** ESI-TOF Mass m/z calcd. $C_{84}H_{98}N_{24}O_{13}$ $[M + 2H]^{2+}$ 825.39; found 826.35

^{1}H NMR (600 MHz, DMSO-d6) δ 10.227 (s, 1H; NH), 10.187 (s, 1H; NH), 10.985 (s, 1H; NH), 9.906 (s, 1H; NH), 9.867 (s, 1H; NH), 9.858 (s, 1H; NH), 9.828 (s, 1H; NH), 9.814 (s, 1H; NH), 8.339 (d, 1H; CH, *J* = 4.5), 8.227 (s, 1H; CH), 8.213 (s, 1H; CH), 8.176 (s, 1H; CH), 8.160 (d, 1H; CH, *J* = 1.8), 8.085 (d, 2H; CH, *J* = 1.5), 8.021–7.996 (m, 4H, CH), 7.944 (m, 1H; CH), 7.89 (m, 1H; CH), 7.877 (m, 1H; CH), 7.505 (s, 1H; CH), 7.459 (s, 1H; CH) 7.288 (s, 1H; CH), 7.225 (s, 1H; CH), 7.213 (s, 1H; CH), 7.197 (s, 1H; CH), 7.133 (m, 2H; CH), 7.055 (s, 1H; CH), 7.027 (s, 1H; CH), 6.887 (br, 2H; NH), 6.846 (br, 3H; NH), 3.941 (s, 3H; N–CH_3), 3.903 (s, 3H; N–CH_3) 3.816 (s, 6H; N–CH_3), 3.808 (s, 6H; N–CH_3), 3.788 (s, 6H; N–CH_3), 3.084 (m, 2H; CH), 2.973 (m, 4H; CH), 2.713 (d, 8H; CH, *J* = 2.7), 2.64 (s, 3H), 2.373–2.316 (m, 6H, CH) 2.174 (m, 6H) 1.961 (m, 2H; CH), 1.786–1.689 (m, 5H; CH), 1.138 (t, 3H; CH)

Polyamide **4** ESI-TOF Mass m/z calcd. $C_{84}H_{98}N_{24}O_{13}$ $[M + 2H]^{2+}$ 825.39; found 825.25

^{1}H NMR (600 MHz, DMSO-d6) δ 10.243 (s, 1H, NH), 10.219 (s, 1H, NH), 10.025 (s, 1H, NH), 10.005 (s, 1H, NH), 9.897 (d, 2H, NH), 9.885 (s, 1H, NH), 9.854 (s, 1H, NH), 8.370 (d, 1H; CH, *J* = 4.2), 8.259 (s, 1H; CH), 8.247 (s, 1H; CH), 8.207 (d, 1H; CH, *J* = 2.04), 8.199 (d, 1H; CH, *J* = 2.1), 8.119 (d, 2H; CH, *J* = 1.5), 8.026–8.055 (m, 4H, CH), 7.978 (m, 1H; CH), 7.922 (d, 1H; CH), 7.858 (m, 1H; CH), 7.540 (d, 2H; CH, *J* = 3), 7.272 (s, 2H; CH), 7.252 (s, 2H; CH), 7.178 (brs, 1H; CH), 7.162 (brs, 1H; CH), 7.149 (brs, 1H; CH), 6.931 (br, 3H; NH), 6.910 (br, 1H; NH), 6.879 (br, 1H; CN), 3.977 (d, 6H; N–CH_3, *J* = 1.8), 3.851 (d, 6H; N–CH_3, *J* = 1.2), 3.827 (d, 6H; N–CH_3, *J* = 1.3), 3.802 (d, 6H; N–CH_3,

$J = 1.6$), 3.107 (m, 2H; CH), 3.010 (m, 2H; CH), 2.746 (d, 6H; CH, $J = 2.4$), 2.413–2.208 (m, 13H, CH) 1.994 (m, 2H; CH), 1.792–1.725 (m, 5H; CH), 1.186 (t, 3H; CH, $J = 7.2$).

4.4.3 Photoreaction

Photoreaction sample was prepared by mixing 10 nM DNA, 10 mM Sodium Cacodylate buffer pH 7.00, 500 mM isopropanol, 100 nM polyamide (0.1% DMSO) and Milli-Q water (total volume 13 μL) and incubated at room temperature for 1 h. For UV irradiation, LED light (model ZUVC30H manufactured by OMRON) with 300 mW at 365 nm was used. The reaction mixture was then treated with 1.25 unit of Uracil-DNA Glycosylase (UDG) and incubated at 37 °C for 1 h. The sample was dried in spin vaccum pump. About 6 μL loading dye was added to the sample and heated at 95 °C for 10 min.

Before loading into the gel, the photoreacted sample was mixed with 6 μL of loading dye (loading dye was prepared by using 300 μl of 0.5 M EDTA, 200 μL of Milli-Q water, 10 ml of formamide, and 2.5 mg of New fuchsin) and heated at 95 °C for 10 min. After cooling down gradually, 1.5 μL of the sample was loaded on a 6% denaturing polyacrylamide gel (7 M Urea). It was then analyzed with sequencing ladder and detected by sequencer (SQ5500E, HITACHI). The sequencing ladder was prepared by Thermosequenase Dye Primer Manual Cycle Sequencing Kit protocol using the template DNA used in this study.

4.4.4 The Methods for SPR Analysis

SPR analysis was performed using BIOCORE X instrument [28, 29]. Biotinylated hairpin DNA was immobilized on streptavidin-coated sensor chip SA to obtain the desired immobilization level (approximately 1200 RU rise after the loading of DNA on the sensor chip). The measurements were carried out using HBS-EP (10 mM HEPES pH 7.4, 150 mM NaCl, 3 mM EDTA, and 0.005% Surfactant P20), purchased from GE Healthcare, with 0.1% DMSO at 25 °C. Different concentrations of polyamides sample were prepared in the HBS-EP buffer with 0.1% DMSO and injected at a flow rate of 20 μL/min. Association rate (k_a), dissociation rate (k_d), and dissociation constant (K_D) were calculated through data processing and fitting with 1:1 binding with mass transfer model using BIAevaluation 4.1 program. The closeness of fit is described by the statistical value χ^2. Specificity was calculated by dividing mismatch/reverse with the match site. All the biotinylated DNA were purchased from JBios and their sequences are as follows:

5′-Biotin-GCGCTTGAGTCGCGTTTTCGCGACTCAAGCGC (PIP1-match)

5′-Biotin-GCGCATGTATCGCGTTTTCGCGATACATGCGC (PIP1-mismatch)

5′-Biotin-GCGCTATGCTCGCGTTTTCGCGAGCATAGCGC (PIP2-reverse)

5′-Biotin-GCGCTTCACTCGCGTTTTCGCGAGTGAAGCGC (PIP1-reverse)

5′-Biotin-GCGCATACGACGCGTTTTCGCGTCGTATGCGC (PIP2-match)

References

1. Willis MC, Hicke BJ, Uhlenbeck OC, Cech TR, Koch TH (1993) Photocrosslinking of 5-iodouracil-substituted RNA and DNA to proteins. Science 262:1255–1257
2. Hicke BJ, Willis MC, Koch TH, Cech TR (1994) Telomeric protein-DNA point contacts identified by photo-cross-linking using 5-bromodeoxyuridine. Biochemistry 33:3364–3373
3. Ogata R, Gilbert W (1977) Contacts between the lac repressor and the thymines in the lac operator. Proc Natl Acad Sci USA 74:4973
4. Krasin F, Hutchinson F (1978) Strand breaks and alkali-labile bonds induced by ultraviolet light in DNA with 5-bromouracil in vivo. Biophys J 24:657–664
5. Suzuki K, Yamauchi M, Oka Y, Suzuki M, Yamashita S (2011) Creating localized DNA double-strand breaks with microirradiation. Nat Protoc 6:134–139. https://doi.org/10.1038/nprot.2010.183
6. Krasin F, Hutchinson F (1978) Double-strand breaks from single photochemical events in DNA containing 5-bromouracil. Biophys J 24:645–656. https://doi.org/10.1016/S0006-3495(78)85410-1
7. Sugiyama H, Tsutsumi Y, Fujimoto K, Saito I (1993) Photoinduced deoxyribose C2′ oxidation in DNA. Alkali-dependent cleavage of erythrose-containing sites via a retroaldol reaction. J Am Chem Soc 115:4443–4448. https://doi.org/10.1021/ja00064a004
8. Sugiyama H, Tsutsumi Y, Saito I (1990) Highly sequence-selective photoreaction of 5-bromouracil-containing deoxyhexanucleotides. J Am Chem Soc 112:6720–6721. https://doi.org/10.1021/ja00174a046
9. Sugiyama H, Fujimoto K, Saito I (1996) Evidence for intrastrand C2′ hydrogen abstraction in photoirradiation of 5-halouracil-containing oligonucleotides by using stereospecifically C2′-deuterated deoxyadenosine. Tetrahedron Lett 37:1805–1808. https://doi.org/10.1016/0040-4039(96)00123-2
10. Watanabe T, Bando T, Xu Y, Tashiro R, Sugiyama H (2005) Efficient generation of 2′-deoxyuridin-5-yl at 5′-(G/C)AA^{X}U^{X}U-3′ (X = Br, I) sequences in duplex DNA under UV irradiation. J Am Chem Soc 127:44–45. https://doi.org/10.1021/ja0454743
11. Watanabe T, Tashiro R, Sugiyama H (2007) Photoreaction at 5′-(G/C)AABrUT-3′ sequence in duplex DNA: efficent generation of uracil-5-yl radical by charge transfer. J Am Chem Soc 129:8163–8168. https://doi.org/10.1021/ja0692736
12. Hashiya F, Saha A, Kizaki S, Li Y, Sugiyama H (2014) Locating the uracil-5-yl radical formed upon photoirradiation of 5-bromouracil-substituted DNA. Nucleic Acids Res 42:13469–13473. https://doi.org/10.1093/nar/gku1133
13. Cook GP, Greenberg MM (1996) A novel mechanism for the formation of direct strand breaks upon anaerobic photolysis of duplex DNA containing 5-Bromodeoxyuridine. J Am Chem Soc 118:10025–10030. https://doi.org/10.1021/ja960652g
14. Cook GP, Chen T, Koppisch AT, Greenberg MM (1999) The effects of secondary structure and O2 on the formation of direct strand breaks upon UV irradiation of 5-bromodeoxyuridine-containing oligonucleotides. Chem Biol 6:451–459
15. Dervan PB (2001) Molecular recognition of DNA by small molecules Bioorg. Med Chem 9:2215–2235
16. Van Dyke MM, Dervan PB (1984) Echinomycin binding sites on DNA. Science 225:1122
17. Dervan PB (1986) Design of sequence-specific DNA-binding molecules. Science 232:464

18. Brenowitz M, Senear DF, Shea MA, Ackers GK (1986) Quantitative DNase footprint titration: a method for studying protein-DNA interactions. Methods Enzymol 130:132
19. Brenowitz M, Senear DF, Shea MA, Ackers GK (1986) "Footprint" titrations yield valid thermodynamic isotherms. Proc Natl Acad Sci USA 83:8462
20. Senear DF, Brenowitz M, Shea MA, Ackers GK (1986) Energetics of cooperative protein-DNA interactions: comparison between quantitative deoxyribonuclease footprint titration and filter binding. Biochemistry 25:7344
21. Trauger JW, Dervan PB (2001) Footprinting methods for analysis of pyrrole-imidazole polyamide/DNA complexes. Methods Enzymol 340:450–466
22. Kaden P, Mayer-Enthart E, Trifonov A, Fiebig T, Wagenknecht HA (2005) Real-time spectroscopic and chemical probing of reductive electron transfer in DNA. Angew Chem Int Ed 44:1636. https://doi.org/10.1002/anie.200462592
23. Wagenknecht HA (2003) Reductive electron transfer and transport of excess electrons in DNA. Angew Chem Int Ed 42:2454. https://doi.org/10.1002/anie.200301629
24. Tashiro R, Ohtsuki A, Sugiyama H (2010) The distance between donor and acceptor affects the proportion of C1′ and C2′ oxidation products of DNA in a BrU-containing excess electron transfer system. J Am Chem Soc 132:14361–14363. https://doi.org/10.1021/ja106184w
25. Netzel TL, Zhao M, Nafisi K, Headrick J, Sigman MS, Eaton BE (1995) Photophysics of 2′-deoxyuridine (dU) nucleosides covalently substituted with either 1-pyrenyl or 1-pyrenoyl: observation of pyrene-to-nucleoside charge-transfer emission in 5-(1-pyrenyl)-dU. J Am Chem Soc 117:9119–9128. https://doi.org/10.1021/ja00141a002
26. Morinaga H, Takenaka T, Hashiya F, Kizaki S, Hashiya K, Bando T, Sugiyama H (2013) Sequence-specific electron injection into DNA from an intermolecular electron donor. Nucleic Acids Res 41:4724–4728. https://doi.org/10.1093/nar/gkt123
27. Meier JL, Yu AS, Korf I, Segal DJ, Dervan PB (2012) Guiding the design of synthetic DNA-binding molecules with massively parallel sequencing. J Am Chem Soc 134:17814–17822. https://doi.org/10.1021/ja308888c
28. Becker HC, Nordén B (1999) DNA binding mode and sequence specificity of piperazinylcarbonyloxyethyl derivatives of anthracene and pyrene. J Am Chem Soc 121:11947–11952. https://doi.org/10.1021/ja991844p
29. BIAevaluation Software Handbook. Version 3 (1999) Biacore AB, Rapsgatan, Sweden

Chapter 5
Examining Cooperative Binding of Sox2 on DC5 Regulatory Element Upon Complex Formation with Pax6 Through Excess Electron Transfer Assay

Abstract Functional cooperativity among transcription factors on regulatory genetic elements is pivotal for milestone decision-making in various cellular processes including mammalian development. However, their molecular interaction during the cooperative binding cannot be precisely understood due to lack of efficient tools for the analyses of protein–DNA interaction in the transcription complex. Here, we demonstrate that photoinduced excess electron transfer assay can be used for analyzing cooperativity of proteins in transcription complex using cooperative binding of Pax6 to Sox2 on the regulatory DNA element (DC5 enhancer) as an example. In this assay, ^{Br}U-labeled DC5 was introduced for the efficient detection of transferred electrons from Sox2 and Pax6 to the DNA, and guanine base in the complementary strand was replaced with hypoxanthine (I) to block intra-strand electron transfer at the Sox2-binding site. By examining DNA cleavage occurred as a result of the electron transfer process, from tryptophan residues of Sox2 and Pax6 to DNA after irradiation at 280 nm, we not only confirmed their binding to DNA but also observed their increased occupancy on DC5 with respect to that of Sox2 and Pax6 alone as a result of their cooperative interaction.

Keywords Excess electron transfer · Transcription factor · Tryptophan
^{Br}U-substituted DNA · Cooperative effect

5.1 Introduction

The complexity of an organ development is manifested through spatiotemporal expression of genes involved in development, which is tightly regulated to a large extent by the combination of transcription factors in multiprotein complexes [1–13]. In these processes, the binding affinity of transcription factors to their genetic elements, which is crucial for transcription activity, is modulated by cooperative binding: low inherent binding affinity of the individual factors is largely enhanced

A. Saha, *Molecular Recognition of DNA Double Helix*, Springer Theses,
https://doi.org/10.1007/978-981-10-8746-2_5

when they present together by their synergistic action. One example that shows this plasticity and diverse combinatorial transcription activity is Sox2, which activates its downstream transcriptional targets by forming cooperative complexes with various factors in each developmental stage. For example, it maintains pluripotency by partnering with various factors like Oct4 and Nanog [5, 7, 11, 12], and controls neurogenesis and retinal developmental by forming complexes with Pax6 [7–10], Otx2 [11], Tlx [12] or Brn2 [13]. Therefore, it is essential to investigate the dynamics of their cooperativity to understand functional intricacies involved in the process of transcription.

Electrophoretic mobility shift assay, widely used for probing the cooperativity and synergistic activity of transcription factors, is often very time-consuming and comes with varying sensitivity. Most importantly, observing the ternary complex between coexisting proteins and DNA is not straightforward because, in many cases, the partner transcription factors interact with low affinity, which limits the possibility to run the intact complex through the gel even at low temperature. In addition, this method cannot provide any detailed molecular mechanism of cooperative binding. Alternatively, high-resolution structural studies such as protein crystallography and NMR can surely give the picture for understanding cooperativity of transcriptional complex, but these methods have limitations in term of sample preparation and technical difficulty. Therefore, there is a need of quick, sensitive, and reproducible alternative method to determine cooperativity of transcription factors in detail. Here, we propose that photoinduced excess electron transfer (EET) from the tryptophan residues of protein to ^{Br}U labeled DNA is an alternative to the classical ways to probe cooperativity by examining the synergistic action of Sox2 and Pax6 on their putative regulatory genetic element called DC5. Transcription factor Pax6 initiates lens development by forming a cooperative complex with Sox2 on the DC5 element, which enhances the lens-specific expression of the δ-crystallin gene (Fig. 5.1a). This specific alliance is responsible for the development of neuronal and retinal tissues [7–10]. For instance, when the Pax6 binding sequence of the DC5 enhancer is replaced with Pax6 binding consensus (DC5con, Fig. 5.1a), the cooperation in binding between Sox2 and Pax6 decreases and the complex failed to activate the reporter gene. Previously, formation of this functional ternary complex was shown by classical EMSA [7, 8] and it was recently analyzed by AFM on a DNA origami frame [14]. Although DNA origami is an attractive platform to observe these complex biological events, the facilities and technical knowledge required to prepare DNA origami frames are available to only a limited number of laboratories. Thus, our proposed method can be used in many cases to detect such crucial biological events based on photosensitive platform.

Long-range EET leads to a great deal of redox chemistry that is widely observed in key biological events such as signaling and sensing [15]. This long-range electron transfer is possible within a distance of 10–25 Å in biological redox reactions [16]. It has been proposed that some repair enzymes use electron transfer from redox cofactors to allow the detection of DNA lesions generated by the oxidation at remote site [17–19]. However, it was also observed that EET can go

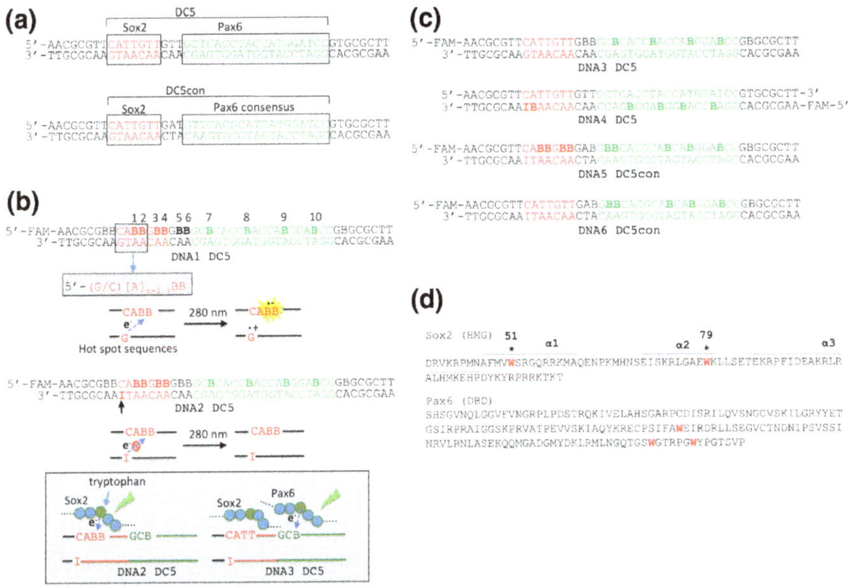

Fig. 5.1 **a** DC5 enhancer element of δ-crystallin gene, in which Sox2 and Pax6 interact during the transcription process leading to the development of neuronal and retinal tissues. The binding sites of Sox2 and Pax6 are shown in red and green, respectively. Also shown the sequence, when the DC5 enhancer is replaced with Pax6 binding consensus (DC5con). **b** DC5 element was modified into DNA1 by replacing thymine residues with its analogous ^{Br}U residues (B represents ^{Br}U) in one strand to capture the excess electrons from the protein. But the motif 5′-CABB in the Sox2-binding site of DNA1 is known for "Hotspot", which is labile for intra-strand electron transfer as shown in the figure. Therefore, DNA1 was further modified to DNA2 by replacing G with hypoxanthine (I) (indicated by an arrow in DNA2) to prevent intra-strand electron transfer. Also shown, the schematic representation of the photoinduced electron transfer from the excited state of tryptophan residues of Sox2 alone (left) and Sox2 with its partner Pax6 to the DC5 DNA element (right). **c** DNA3 was designed to capture an electron from Pax6 by replacing "T" with "^{Br}U" in the Pax6 binding sites as shown in Fig. 1b (right). DNA4 was designed to check the photoreactivity of the complementary strand by incorporating ^{Br}U residues. DNA5 and DNA6 were designed based on DC5con sequence. **d** The amino acid sequences of Sox2(HMG) and Pax6 (DBD) are shown. Trp residues are shown in red

through without redox cofactors as evidenced in the case of *Escherichia coli* DNA photolyase that can repair thymine dimers without the aid of redox cofactors [20, 21]. Moreover, it was also reported that thymine dimers can be repaired by Lys-Trp-Lys motif under irradiation conditions [22–25]. These results together suggest that specific amino acids are responsible for the repair process through EET. Requirement of specific amino acids as electron donor was further supported in a previous report, where it was observed that the photoreactivity of DNA can be achieved by the photoinduced single-electron transfer from Trp residue of the DNA-binding protein using an electron acceptor ^{Br}U base [25, 26]. Based on these results, we hypothesized that photoinduced electron transfer from proteins to DNA could be applied to check the specific interaction between protein and DNA, and can be further utilized to probe cooperativity of proteins in transcription complex. To test this possibility, we applied our strategy to detect protein–DNA interaction conducted by Sox2 or Pax6 alone or in their complex. We modified DC5 element and DC5 Con by replacing thymine with its ^{Br}U analog at the binding site of the protein. ^{Br}U is an attractive synthetic nucleotide as its substitution does not affect the functionality of the resulting DNA but can easily trap an electron during DNA-mediated EET. The trapped excess electron converts ^{Br}U to a uracil-5-yl radical (U$^{\bullet}$) by eliminating the bromide ion [27]. As a result, EET from protein to ^{Br}U results in the strand cleavage by the heat treatment since the heat-labile 2-deoxyribonolactone is generated from U$^{\bullet}$ radical by intra-strand hydrogen abstraction from the deoxyribose moiety of the 5′-side at the C1′ and C2α′ position, respectively [28, 29]. The strand cleavage can be accelerated further by including isopropanol (^{i}PrOH) as an excess H-atom source and subsequent treatment of uracil-DNA glycosylase (UDG) as previously reported [30, 31]. Under this strategy, Sox2- or Pax6 and their cooperative binding to DNA can be monitored by the strand cleavage due to trapping of excess electrons from tryptophan residues to ^{Br}U. For this purpose, we used purified recombinant DNA-binding domains: Sox2 (HMG), HMG representing high-mobility group, and Pax6(DBD), DBD representing DNA-binding domain.

5.2 Results and Discussions

5.2.1 Designing DC5 by Incorporating ^{Br}U and Hypoxanthine (I) to Capture an Electron from Sox2(HMG) or Pax6(DBD)

To test the aforementioned possibility, we applied our strategy to detect protein–DNA interaction conducted by Sox2 or Pax6 alone or in their complex. We modified DC5 element and DC5 Con by replacing thymine with its ^{Br}U analog at the binding site of the protein (as shown in Fig. 5.1b, c). Under this strategy, Sox2- or Pax6 and their cooperative binding to DNA can be monitored by the strand cleavage due to trapping of excess electrons from tryptophan residues to ^{Br}U. It is to be remembered that as a result of EET from protein to ^{Br}U, can generate the reactive uracil-5-yl radical (U$^{\bullet}$) giving strand cleavage by H-atom abstraction from nearest sugar moiety. For this purpose, we used purified recombinant DNA-binding domains: Sox2(HMG), HMG representing high-mobility group, and Pax6(DBD), DBD representing DNA-binding domain (Fig. 5.1d).

We first designed DNA1, a double strand DC5 element containing ^{Br}U residues at the binding site of Sox2 and Pax6 (Fig. 5.1b). We expected that DNA1 would capture electrons at the respective protein-binding sites due to close proximity of any of the tryptophan residues of the binding protein when incubated with proteins under UV irradiation (Fig. 5.1d). But it is known that specific sequences in the ^{Br}U-substituted DNA, for instance, 5′-G/C[A]$_{n\ =\ 1,2,3}$^{Br}U^{Br}U-3′ sequences, called hotspots can induce intra-strand electron transfer under UV irradiation condition [33, 34]. Since G has the lowest oxidation potential among the four bases (The oxidation potentials of G, A, T, and C are 1.24, 1.69, 1.9, and 1.9 V, respectively [32]), an electron from G transfers to ^{Br}U^{Br}U residues through intervening A bridges adjacent to ^{Br}U (as shown in the Fig. 5.1b) and A bridges in this case helps in forward and backward electron transfer process [33, 34]. A same hotspot sequence, CA^{Br}U^{Br}U was found in the Sox2 binding sequence while in the Pax6 binding site (in both cases DC5 and DC5con) no such sequences were noticed (even though there are many G's but lacks A bridges between G and ^{Br}U^{Br}U). To quench this intra-strand electron transfer from G to ^{Br}U ^{Br}U in the Sox2 binding site of DNA1 and to eliminate unintended false positives, we constructed DNA2 (Fig. 5.1b) in which G was replaced by hypoxanthine (I), a modified purine base, in its complementary strand (indicated by an arrow) only in the Sox2 site. We hypothesized that this modification would prevent any intra-strand electron transfer due to the higher oxidation potential of I than that of G (1.4 vs. 1.24 V) as depicted in a schematic representation shown in Fig. 5.1b [32]. To test this principle, we irradiated DNA1 and DNA2 at 280 nm for 30 min and analyzed the resulting samples by denaturing gel electrophoresis (Fig. 5.2).

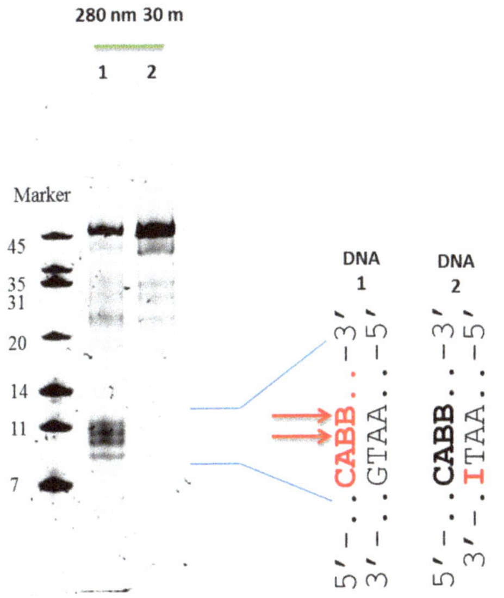

Fig. 5.2 Deactivation of hotspot sequence in the DC5 enhancer, in which Sox2 and Pax6 cooperatively interact during transcription process. DNA1 contains the hotspot sequence shown in red, whereas in DNA2 the same hotspot sequences were deactivated by replacing G with I shown in red. DNA1 and 2 were irradiated using 280 nm for 30 min and after subsequent UDG treatment, it was then analyzed by denaturing PAGE (20%). Distinct cleavage band appeared in the hotspot sequence of DNA1 (lane1), whereas almost no cleavage appeared in the hotspot sequence of DNA2 (lane 2). The concentration of DNA in all samples was 1.25 μM

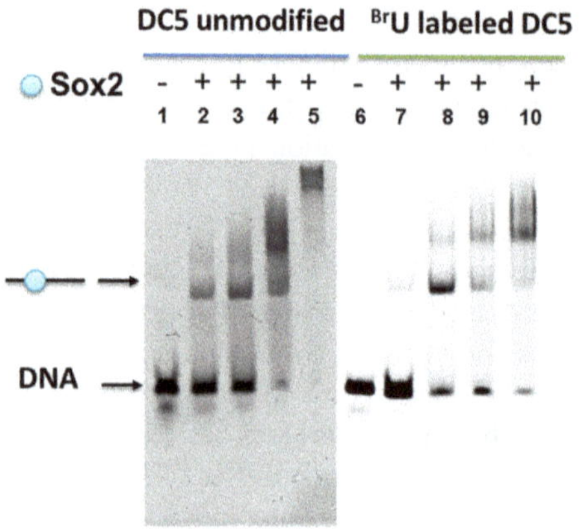

Fig. 5.3 The EMSA (5% polyacrylamide native gel) for Sox2 protein is shown in the unmodified DC5 (left) and ^{Br}U labeled DC5 (right), DNA2. With the increase in concentration of Sox2, clear gel shift was observed. Lane 1 and 6, DNA; lane 2 and 7, DNA2 + Sox2 (1.25 μM); lane 3 and 8, DNA2 + Sox2 (2.5 μM), lane 4 and 9, DNA2 + Sox2 (6.25 μM); lane 5 and 10, DNA2 + Sox2 (12.5 μM). In all sample, 1.25 μM DNA was used. The sample was prepared in a buffer containing 10 mM Tris-HCl pH 7.5, 1 mM EDTA, 50 mM KCl, 100 μg/mL BSA, 5% v/v glycerol, and before loading into gel it was incubated at 4 °C for 2 h

While DNA1 got cleaved, almost no cleavage was observed in DNA2, suggesting the absence of intra-strand electron transfer. Therefore, this method was employed in other assays to eliminate any false readout in the hotspot. This exercise was not necessary for Pax6 binding site due to lack of similar hotspots (Fig. 5.1a). To examine whether DNA2 limits DNA binding capacity of Sox2, we performed EMSA using Sox2(HMG) on DNA2 and compared with unmodified DC5. We observed Sox2(HMG) binds to DNA2 in almost similar stoichiometric ratio to unmodified DC5 (Fig. 5.3).

5.2.2 *Capturing an Electron from Sox2(HMG) Upon Binding to DNA*

We checked the photoreactivity of DNA2 upon treatment of Sox2(HMG) at a molar ratio of 2:1 and 5:1 (protein: DNA) under irradiation at 280 nm for 15 min at 0 °C in a buffer containing Tris-HCl pH 7.5 and 200 mM ^{i}PrOH. After irradiation, the sample was subjected to UDG treatment for converting U into the heat-labile abasic site (AP site). When the sample was heat treated at 95 °C for cleavage and analyzed by 20% denaturing polyacrylamide gel electrophoresis (PAGE), two distinct DNA fragments were observed (lane 5 in Fig. 5.4). Based on the size of the fragments,

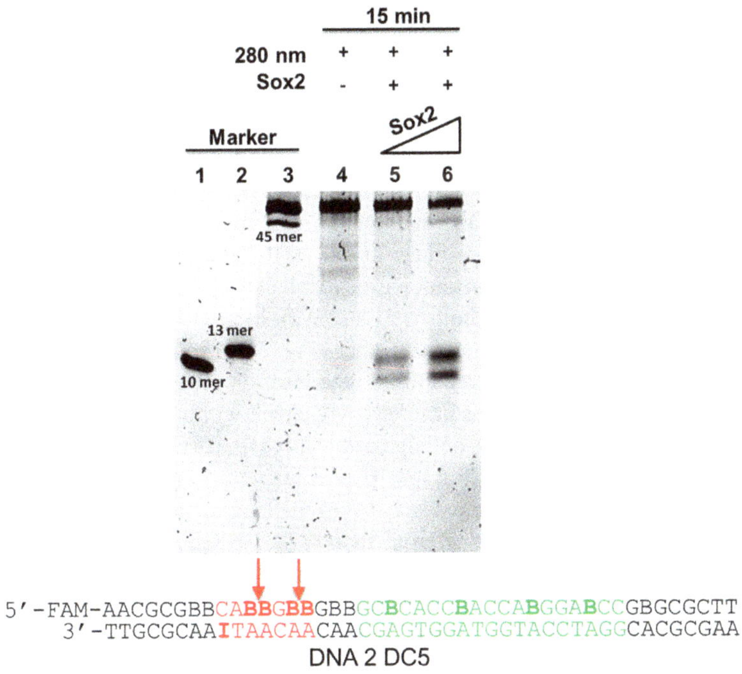

Fig. 5.4 Analyses of the DNA cleavage as a result of photoinduced electron transfer from Sox2 (HMG) to the modified DC5 element (DNA2) using 20% polyacrylamide denaturing gel (7 M urea). The gel electrophoresis was performed at 200 V and 250 mA for 140 min and checked by Fujifilm (FLA-3000). Lanes 1, 2, and 3 are DNA size markers with sizes of 45, 10, and 13 mer, respectively; Lane 4, DNA2 with UV irradiation at 280 nm for 15 min; lane 5, UV irradiation at 280 nm for 15 min on Sox2(HMG): DNA2 at 2.5 μM:1.25 μM ratio; lane 6, UV irradiation at 280 nm for 15 min on Sox2(HMG): DNA2 at 6.25 μM:1.25 μM ratio. Concentration of DNA was 1.25 μM. Reaction was performed in a buffer containing 10 mM Tris-HCl pH 7.5, 1 mM EDTA, 50 mM KCl, 100 μg/mL BSA, 5% v/v glycerol, and 200 mM ^{i}PrOH. Each sample was incubated with 1.25 U of UDG at 37 °C for 1 h after irradiation and finally heated at 95 °C for 10 min

they were expected to be the cleavage products at the Sox2-binding site (indicated by arrows in Fig. 5.4). This interpretation is further confirmed by the increased band intensity when DNA2 was incubated with the higher concentration of Sox2 (HMG) (lane 6 in Fig. 5.4). Interestingly, no DNA cleave was detected other than at the Sox2-binding site, manifesting that Sox2 binding to the DC5 element is sequence-specific. It should be noted that there was no strand cleavage in absence of Sox2(HMG) under the same reaction condition (lane 4 in Fig. 5.4). Taken together, it can be concluded that the photoinduced electron transfer from protein to DNA and subsequent cleavage of DNA can be used for investigating sequence-specific binding of transcription factors to their corresponding DNA elements.

5.2.3 Locating the Tryptophan Residues of Sox2(HMG)

To get a clear view of placement of electron donor–acceptor couples within the protein–DNA complex, a model of human Sox2(HMG) bound to 5′-CA^{Br}U^{Br}UG^{Br}U^{Br}U-3′/5′-GTAACAA-3′ was constructed by placing the hSox2 (HMG) to the DC5 element (5′-CATTGTT-3′/5′-GTAACAA-3′) after replacing thymine with ^{Br}U (see materials and methods). The complex model reveals that the first two helices comprising 45 residues in the HMG domain are involved in the sequence-specific binding to the minor groove of DNA (Fig. 5.5). Inspection of the crystal structure inferred that both W51 and W79 located near the DNA–protein interface are in close proximity to each other (Fig. 5.5a). As a result, tryptophan residues, W51 and W79, are located close to the electron acceptor ^{Br}U (B1 and B2) at distances of 17.5 and 13.0 Å, respectively, which are within the range of distance for potent electron transfer. We anticipate that both the tryptophan residues are putative electron donors but W79 might play a major role in the photoreactivity considering their orientation and distance. From the surface representation of Sox2 (HMG) in the model, it can be inferred that W79 located on the DNA–protein interface is in close proximity to the electron acceptor ^{Br}U (B1 and B2) at distances of 13 and 11 Å, respectively (Fig. 5.5b). In addition, W79 seems to also transfer the electrons to B3 and B4 as well, since distances from indole nitrogen of W79 to bromine are 10 and 14 Å, respectively (Fig. 5.5b). However, W51 appears to be located on the surface opposite to that of protein–DNA interface and thus it might have less electron transfer propensity.

5.2.4 Capturing an Electron from Pax6(DBD) Upon Complex Formation with Sox2(HMG) and DNA3

We then designed DNA3 to capture the transferred electrons from Pax6 by substituting thymine with ^{Br}U in the Pax6 binding region (Fig. 5.1c). The photoreactivity of DNA3 was examined similarly in the presence of either Pax6(DBD) alone or in combination of Pax6(DBD) with Sox2(HMG) (Fig. 5.6). It is well known that most of the transcription co-activators of Sox2 have low inherent DNA binding affinity in the absence of Sox2 but gain affinity when they form a ternary complex in the presence of Sox2. As expected, Pax6(DBD) alone was not able to induce any DNA cleavage, representing that electrons were not transferred from the tryptophan residues of Pax6 to DNA, possibly due to the transient binding of protein to DNA (lane 5 in Fig. 5.6). However, in the presence of both Sox2(HMG) and Pax6(DBD), electron transfer from Pax6 led to strand cleavage (lane 6 in Fig. 5.6). Pax6(DBD)

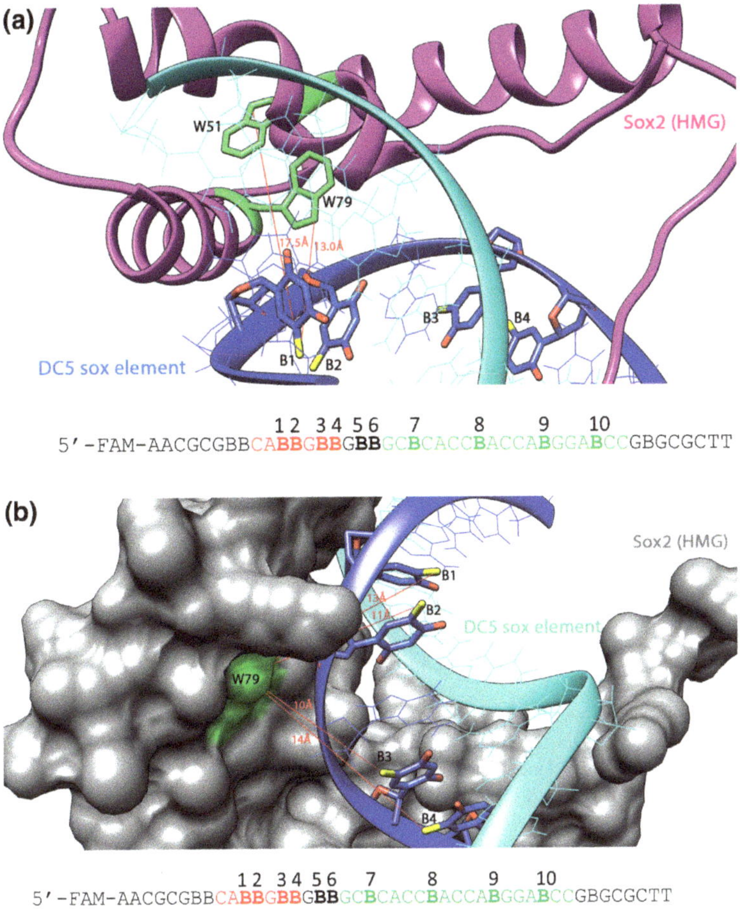

Fig. 5.5 **a** Ribbon diagram of hSox2(HMG) bound to 5′-CAB1B2GB3B4-3′/5′-AACAATG-3′, which was constructed based on the structure of human Sox2(HMG) and structure of DC5 Sox2 DNA element as shown below. B base represents ^{Br}U. Tryptophan residues (W51 and W79) and ^{Br}U nucleosides (B1–B4) are drawn as green and blue stick models, respectively, and Br and oxygen atoms in ^{Br}U are shown in yellow and red, respectively. **b** Surface representation of hSox2 (HMG) bound to 5′-CAB1B2GB3B4-3′/5′-AACAATG-3′. B represents ^{Br}U. Tryptophan residue (W79) and ^{Br}U nucleotides (B1–B4) are colored in green and blue, respectively, and Br and oxygen atoms in ^{Br}U are shown in yellow and red, respectively. The sequence is shown below

contains three Trp residues, thus it cannot be concluded that which residues are involved in electron transfer due to the lack of high-resolution model of Pax6/DC5 complex. However, it is obvious that one of Trp residues of Pax6 approaches to proximal position to ^{Br}U by forming a stable complex with DNA due to the

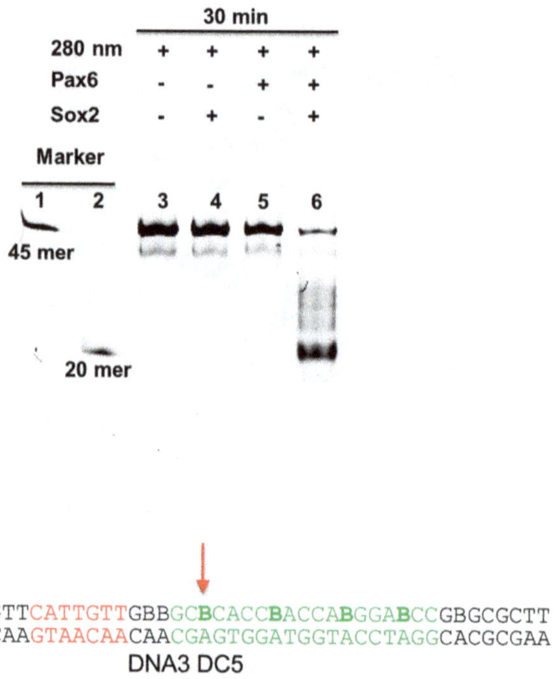

Fig. 5.6 Analyses of DNA cleavage by photoinduced electron transfer from Pax6(DBD) to DNA3 in the Sox2-Pax6-DNA3 complex using 20% polyacrylamide denaturing gel (7 M urea). Lane 1, 45 mer DNA marker; lane 2, 20 mer DNA marker; lane 3, UV irradiated DNA3 at 280 nm for 30 min; lane 4, UV irradiated DNA3 with 6.25 μM of Sox2, lane 5, UV irradiated DNA3 with 6.25 μM of Pax6; lane 6, UV irradiated DNA3 with 6.25 μM of Sox2 and Pax6; The concentration of DNA3 was 1.25 μM in all samples

conformation change accompanied by Sox2 binding. The cooperative binding of Sox2–Pax6 on DNA3 and increased DNA binding affinity of Pax6 were also confirmed by a conventional EMSA (Fig. 5.7), which validated the result of EET assay. It is to be noted that photoreaction by these proteins on the opposite strand containing ^{Br}U, DNA4 doesnot have any excess electron transfer. It suggests that the putative electron donor are far from this strand.

5.2.5 *Validating the Influence of Pax6(DBD) on Sox2 (HMG) Binding by Electron Transfer*

It has been demonstrated that Pax6 enhances the binding affinity of Sox2 to DC5 by AFM on a DNA origami frame [14]. To confirm the enhanced binding affinity of Sox2 upon cooperative binding of Pax6 using EET assay, we investigated the

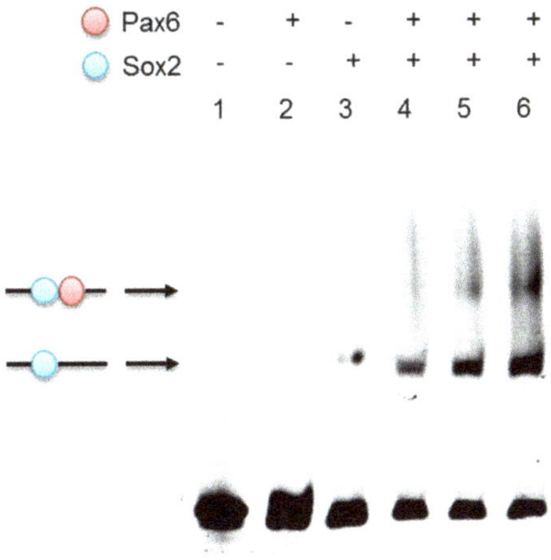

Fig. 5.7 EMSA (5% polyacrylamide native gel) is shown to confirm the cooperativity between Sox2-Pax6 and DNA3. Lane 1, DNA3; lane 2, DNA3 + Pax6 (12.5 μM); lane 3, DNA3 + Sox2 (2.5 μM); lane 4, DNA3 + Sox2 (2.5 μM) + Pax6 (2.5 μM), lane 5, DNA3 + Sox2 (2.5 μM) + Pax6 (6.25 μM); lane 6, DNA3 + Sox2 (2.5 μM) + Pax6 (12.5 μM). The sample was prepared in a buffer containing 10 mM Tris-HCl pH 7.5, 1 mM EDTA, 50 mM KCl, 100 μg/mL BSA, 5% v/v glycerol, and before loading into gel it was incubated at 4 °C for 12 h

electron transfer from Sox2 to DNA2 in the presence and absence of Pax6 under irradiation conditions by examining the DNA cleavage (Fig. 5.8). DNA2 was incubated with Sox2(HMG) alone or with Sox2(HMG)/Pax6(DBD) and irradiated for 15 min following the similar photoreaction procedure, and the amount of the cleavage DNA was compared. To assess the extent of any synergistic effect, we added a minimum amount of Pax6 (0.5 equiv.) to a sample containing two equivalent Sox2, thus to ensure that at this amount it should not cleave the DNA at its binding site. Consistently with the previous report confirmed by AFM, the DNA binding affinity of Sox2 was enhanced about two folds by Pax6 binding as manifested by the increased cleavage content from 42.1 to 82.5% before and after Pax6 binding, respectively. It is to be noted that Pax6 alone (lane 2) at 2 equiv. was unable to give any cleavage. This result suggests that, like Sox2, Pax6 also plays an important role in stabilizing the protein/DNA complex. Given that strand cleavage through electron transfer from Sox2–Pax6 to DNA is more extensive than that from Sox2 or Pax6 alone, it is clear that a cooperative binding partnership is in operation.

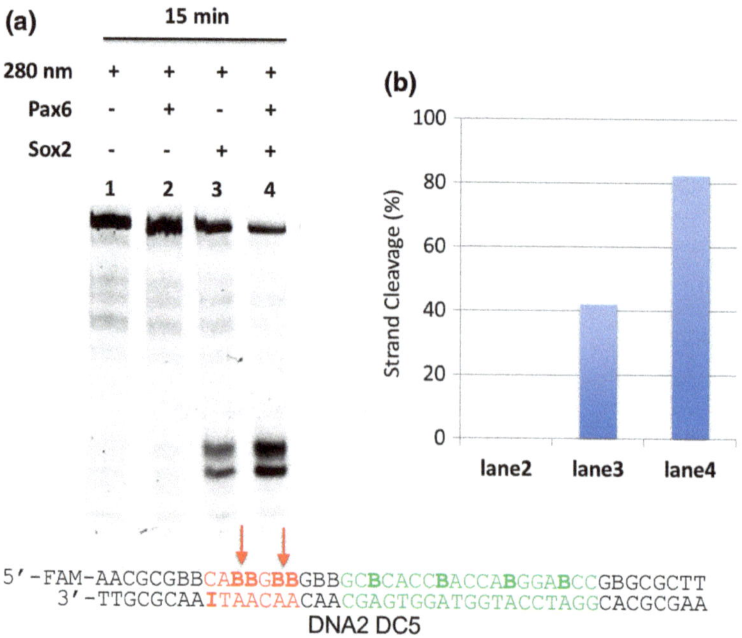

Fig. 5.8 **a** The analyses of the DNA cleavage by photoelectron transfer from Sox2 to DNA in the presence of Pax6. The photoreaction procedure is described in Fig. 5.3. Lane 1, DNA2 irradiated by UV 280 nm for 15 min; lane 2, Pax6(DBD) (2.5 μM) in lane 1 condition; lane 3, Sox2(HMG) (2.5 μM) in the lane 1 condition; Lane 3, Sox2(HMG) (2.5 μM) and Pax6(DBD) (0.625 μM) in the lane 1 condition. The concentration of DNA2 was 1.25 μM in all samples. **b** Quantification of the extent of strand cleavage shown in lanes 2, 3 and 4

5.2.6 Electron Transfer from Pax6(DBD) on DC5con: Determining the Structure–Function Relationship Between DC5 and DC5con

Because it has been found that DNA molecule commonly plays an active role in cooperative interactions, thus we were interested to see whether EET can help in finding the structure–function relationship or not [35]. One such example is the replacement of Pax6 binding sequence from the DC5 enhancer with Pax6 binding consensus as shown in Fig. 5.1a, results in the disruption of cooperation between Pax6 and Sox2. As a result the complex could not activate the reporter gene [7]. In order to probe this crucial structure–function relationship using EET we modified the DC5con by DNA5 and DNA6 (Fig. 5.1a, c). We designed DNA5 by replacing thymine with ^{Br}U in the DC5con sequence to confirm EET by Sox2. Our results revealed that Sox2 alone was able to cleave DNA5 as shown in Fig. 5.9 (lane 6). Since there is a possibility that Sox2 may degrade by the electron injection (oxidation of tryptophan) to DNA5, thus it may disturb the formation of ternary

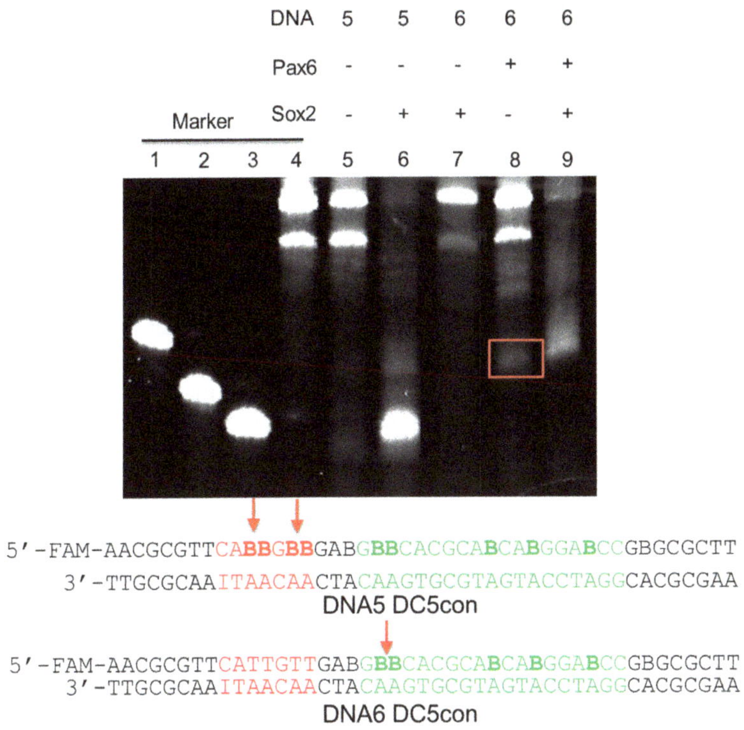

Fig. 5.9 Analyses of DNA cleavage by photoinduced electron transfer from Sox2(HMG) and Pax6(DBD) to DC5Con sequence (DNA5–6) using 15% TBE-Urea Gel (Invitrogen), 180 V, 90 min. Lane 1, 20 mer DNA marker; lane 2, 14 mer DNA marker; lane 3, 11 mer DNA marker; lane 4, 45 mer marker; lane 5, UV irradiated DNA5 at 280 nm for 30 min; lane 6, UV irradiated DNA5 with 6.25 μM of Sox2; lane 7, UV irradiated DNA6 with 6.25 μM of Pax6; lane 8, UV irradiated DNA6 with 6.25 μM of Sox2 and Pax6. The concentration of DNA was 1.25 μM in all samples. Asterisk represents an impurity and it is not generated by photoreaction (because it is seen in control lane 4)

complex. That is why we decided to design DNA6 in which thymine was replaced with ^{Br}U only in the Pax6 binding site. A similar strategy was applied with DNA2 (DC5 sequence) which was used for Sox2 (Fig. 5.4), while a different DNA3 was used for Pax6 induced EET (Fig. 5.6). Interestingly in the DC5con, we revealed that Pax6 alone cleaved the DNA6 at 17.5% at its binding site (Fig. 5.9, lane 8), while in case of DC5 no such cleavage was seen (Fig. 5.6, lane 5). When we added Sox2 and Pax6 with DNA6 we observed that it induces strand cleavage higher than Pax6 alone (lane 9), which indicates that still the cooperativity is active but less compare to DC5.

5.2.7 *Examining the Fate of Tryptophan Residues After Electron Injection*

Finally, we examined the fate of the tryptophan residues of Sox2 after electron injection into the DNA. Previously, we reported that the fluorescence intensity of tryptophan in Sso7d was quenched after electron transfer to DNA [26]. We were thus interested to know the fate of the tryptophan residues of both proteins after irradiation at 280 nm. Upon excitation at 295 nm, both Sox2(HMG) and Pax6 (DBD) release a characteristic emission peak from tryptophan at 330 and 350 nm, respectively. Interestingly, a significant loss of fluorescence intensity of tryptophan residues of Sox2(HMG) was observed after irradiation for 15 min, as shown in Fig. 5.10a. The loss of emission at 330 nm indicates the photodegradation in tryptophan residues. In contrast, no significant loss of fluorescence intensity, except a little quenching caused by the direct binding was observed in the tryptophan residues of Sox2(HMG) after irradiating the DC5 DNA element that has no ^{Br}U residues. The fluorescent intensity of tryptophan was recovered after adding a high concentration of NaCl (2.0 M) into the reaction mixture. This result suggests that electron transfer occurs from Sox2(HMG) to the ^{Br}U residues of DNA2. In contrast, the intensity of fluorescence of Pax6 with DNA2 was not affected even after 30 min irradiation, confirming that no electron transfer from the protein occurs due to the weak binding affinity of Pax6 to DNA2 (Fig. 5.10b).

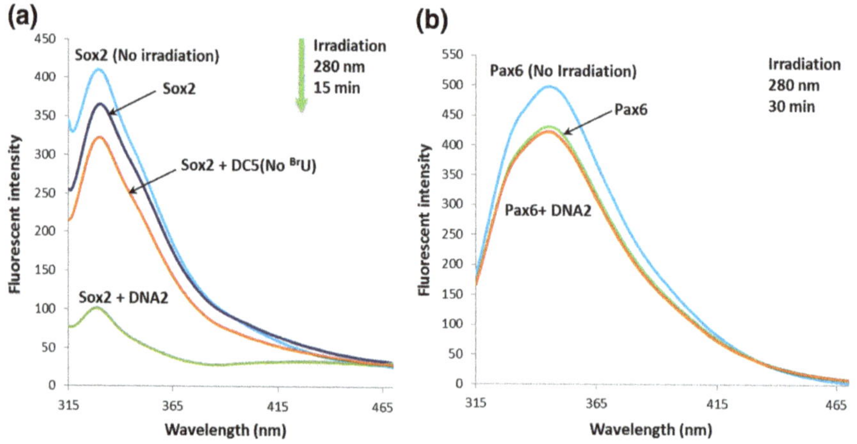

Fig. 5.10 **a** Fluorescence intensity of tryptophan residues of Sox2. Fluorescence emission spectra of Sox2 before irradiation (blue) and after irradiation (dark blue) at 280 nm for 15 min. Emission spectra of Sox2 incubated with DNA2 (green) and with DC5 (orange), after irradiation for 15 min (excitation at 295 nm, 10 nm slit width). **b** Fluorescence emission spectra of Pax6 alone before irradiation (blue), after irradiation (green) and Pax6 when incubated with DNA2 irradiated at 280 nm for 30 min (orange) (excitation at 295 nm, 5 nm slit width). Photoreaction procedure for Sox2 and Pax6: reaction mixture was prepared using 20 mM sodium cacodylate buffer pH 7.50 and 2 μM of each protein and DNA in 50 μL (final volume), irradiated at 0 °C

5.3 Conclusions

To summarize, we demonstrate the characteristic features of cooperativity of Sox2 and its partner Pax6 during the formation of a functional ternary complex with DC5 DNA element by examining electron transfer processes. This approach involved selected substitution of thymine with ^{Br}U residues at the protein-binding sites on the DC5 element and confirmation of binding by the cleavage of the ^{Br}U-substituted DNA. Since Trp is frequently involved in protein–DNA interaction, Trp-induced EET assay can be a powerful tool for investigating protein–DNA interaction and possibly applied to many cases. To our knowledge, this is the first time to observe a crucial biological event, such as cooperative binding of transcription factors to DNA using the EET assay. This study opens up ample opportunities to use EET as an efficient photosensitive method for probing DNA–protein interactions. Consequently, the application repertoire of this method can be expanded to various important biological events accompanied by a dynamic assembly of macromolecular complexes.

5.4 Materials and Methods

5.4.1 Preparation of DNA

All the oligos were purchased from Jbios and Sigma Genosys and used without further purification.

5.4.2 Preparation of Sox2 and Pax6

These two proteins were amplified in a similar way as before [14]. The gene encoding the HMG domain (residues, 1–117) of human Sox2 was amplified and cloned in pVFT1S using EcoR1 and Xho1 sites and, the gene encoding the DNA-binding domain (residues 4–169) of human Pax6 was amplified and cloned into pET41a (Novagen, Darmstadt, Germany) using Nco1 and Xho1 sites. *E. coli* BL21 (DE3) Star (Invitrogen, CA, USA) transformed with pVFT1S-Sox2(HMG) and pET41a-Pax6(DBD) were grown in LB media containing Kanamycin (Duchefa Biochemie, Netherlands) at a concentration of 50 μg/ml. Both cultures were grown for 12 h and then diluted 1:100 into fresh LB media and incubated at 37 °C by shaking until the OD 600 was 1.0. Cells were then induced with a final concentration of 0.5 mM isopropyl-β-D-thiogalactopyranoside (IPTG) (EMD Chemicals, CA, USA) and incubated for 6 h. Cells were harvested by centrifugation at 5000 rpm for 15 min at 4 °C and stored at −80 °C until used. Cell pellets expressing Sox2(HMG) were suspended in buffer A (25 mM Tris-Cl, pH 8.0,

500 mM NaCl, 40 mM imidazole, and 1 mM PMSF) and disrupted with a vibra cell sonicator (Sonics & Materials Inc., CN, USA) using a pulse of 2 s ON and 4 s OFF for a total of 10 min at an amplitude of 60%. Cell lysate was clarified by centrifugation at 30,000 g for 90 min at 4 °C. The supernatant was applied to a nickel-charged NTA column (GE Healthcare, NJ, USA), pre-equilibrated with buffer B (25 mM Tris-HCl, pH 8.0, 500 mM NaCl, and 40 mM imidazole). The column was washed with ten column volumes (CV) of buffer B, and the protein was eluted with a gradient from 40 mM to 1 M imidazole in 40 CV. The fractions containing Sox2(HMG) were concentrated to less than 5 ml using a centrifuge filtration device Vivaspin 20 (Sartorius Stedim Biotech, Goettingen, Germany) and then applied to a Superdex 200 column (GE Healthcare, NJ, USA) equilibrated with buffer C (25 mM Tris–HCl, pH 8.0, 500 mM NaCl, and 1 mM DTT). Cell pellets expressing GST-Pax6(DBD) were suspended in buffer D (25 mM Tris-HCl, pH 7.5, 500 mM NaCl, 40 mM imidazole, and 1 mM PMSF) and disrupted with a vibra cell sonicator (Sonics & Materials Inc., CN, USA) using a pulse of 2 s ON and 4 s OFF for a total of 10 min at an amplitude of 60%. The cell lysate was clarified by centrifugation at 30,000 g for 90 min at 4 °C. The supernatant was applied to a nickel-charged NTA column (GE Healthcare, NJ, USA) pre-equilibrated with buffer D. The column was washed with 10 CV of buffer B, and the protein was eluted with a gradient from 40 mM to 1 M imidazole in 50 CV. The fractions containing Pax6(DBD) were concentrated using a centrifuge filtration device Vivaspin 20 (Sartorius Stedim Biotech, Goettingen, Germany). The proteins were then moved to buffer E (25 mM Tris-HCl, pH 7.5, 500 mM NaCl, and 1 mM DTT).

5.4.3 *Irradiation Set Up*

HM-3 hypermonochromator (Jasco) was used for irradiation at 280 nm. The Eppendorf tube containing the sample (DNA and protein) was attached vertically with the light fiber using a holder so that the light can pass through the sample and placed in an ice box. The description of the machine used in this study can be found at [36].

5.4.4 *Photoreaction Scheme*

DNA and protein were mixed in a buffer containing 10 mM Tris-HCl pH 7.5, 1 mM EDTA, 50 mM KCl, 100 μg/mL BSA, 5% v/v glycerol, and 200 mM ^{i}PrOH. The sample was incubated at 4 °C for minimum 1 h and then irradiated using 280 nm UV transilluminator. The sample was treated with UDG enzyme

(1.25 U), incubated at 37 °C for 1 h. After the enzymatic digestion the sample was dried up completely using high vacuum pump and to it about 10 μL of loading dye (containing of 300 μl of 0.5 M EDTA, 200 μl of Milli-Q water, 10 ml of formamide, and 2.5 mg of new fuchsin) were added and finally heated at 95 °C for 10 min. The sample was then analyzed by denaturing PAGE (20%).

5.4.5 Fluorescence Measurements

Steady state fluorescence measurements of photoirradiated Sox2 and Pax6 samples were conducted with a Jasco FP-6300 spectrofluorometer. Measurements were performed by using a fluorescence cell with a 0.5-cm path length. Fluorescent intensity of Sox2 was quenched when it forms a complex with DNA2, it was difficult to estimate the conversion of the Trp in Sox2. To overcome this trouble, we added a high concentration of NaCl (2 M final concentration) to the reaction mixture after the photoirradiation and checked the fluorescent intensity after keeping the solution in ice for at least 30 min. Gradually, the fluorescence intensity of the Trp is recovered (~90–95%) because Sox2 cannot bind to DNA in such high-salt conditions. The fluorescence intensity of the Trp of Sox2 was not affected in these conditions. A similar condition was used for Pax6.

5.4.6 Page

After photoreaction cleaved DNA was analyzed on a 20% polyacrylamide denaturing gel (7 M Urea). The condition of electrophoresis was 200 V and 250 mA for 140 min. For **EMSA**, the electrophoresis condition was 80 V and 140 mA for 80 min. The gel was analyzed by FLA-3000 (Fujifilm) and the cleavage amount was measured by using multigauge v3.1 software.

5.4.7 Model Building

The model of hSox2(HMG) bound to DC5 element was made using coordinates of hSox2(HMG) in the structure of hSox2/FGF4 (PDB ID: 1GT0) and the DNA corresponding to DC5 Sox2 DNA element in the structure of SRY(HMG)/DC5 (PDB ID: 2GZK) by superposing the hSox2 (HMG) on SRY in 2GZK followed by manual modeling and energy minimization to get a pose of possible interaction of hSox2(HMG) with DC5 Sox2 DNA element.

References

1. Levine M, Tjian R (2003) Transcription regulation and animal diversity. Nature 424:147–151. https://doi.org/10.1038/nature01763
2. Vaquerizas JM, Kummerfeld SK, Teichmann SA, Luscombe NM (2009) A census of human transcription factors: function, expression and evolution. Nat Rev Genet 10:252–263. https://doi.org/10.1038/nrg2538
3. Petti AA, McIsaac RS, Ho-Shing O, Bussemaker HJ, Botstein D (2012) Combinatorial control of diverse metabolic and physiological functions by transcriptional regulators of the yeast sulfur assimilation pathway. Mol Biol Cell 23:3008–3024. https://doi.org/10.1091/mbc.E12-03-0233
4. Cunha PMF, Sandmann T, Gustafson EH, Ciglar L, Eichenlaub MP, Furlong EEM (2010) Combinatorial binding leads to diverse regulatory responses: Lmd is a tissue-specific modulator of Mef2 activity. PLoS Genet 6:e1001014. https://doi.org/10.1371/journal.pgen.1001014
5. Reményi A, Lins K, Nissen LJ, Reinbold R, Schöler HR, Wilmanns M (2003) Crystal structure of a POU/HMG/DNA ternary complex suggests differential assembly of Oct4 and Sox2 on two enhancers. Genes Dev 17:2048–2059. https://doi.org/10.1101/gad.269303
6. Ng CK, Li NX, Chee S, Prabhakar S, Kolatkar PR, Jauch R (2012) Deciphering the Sox-Oct partner code by quantitative cooperativity measurements. Nucleic Acids Res 40:4933. https://doi.org/10.1093/nar/gks153
7. Kamachi Y, Uchikawa M, Tanouchi A, Sekido R, Kondoh H (2001) Pax6 and SOX2 form a co-DNA-binding partner complex that regulates initiation of lens development. Genes Dev 15:1272–1286. https://doi.org/10.1101/gad.887101
8. Narasimhan K, Pillay S, Huang YH, Jayabal S, Udayasuryan B, Veerapandian V, Kolatkar P, Cojocaru V, Pervushin K, Jauch R (2015) DNA-mediated cooperativity facilitates the co-selection of cryptic enhancer sequences by SOX2 and PAX6 transcription factors. Nucleic Acids Res 43:1513. https://doi.org/10.1093/nar/gku1390
9. Morimura H, Tanaka S, Ishitobi H, Mikami T, Kamachi Y, Kondoh H, Inouye Y (2013) Nano-analysis of DNA conformation changes induced by transcription factor complex binding using plasmonic nanodimers. ACS Nano 7:10733. https://doi.org/10.1021/nn403625s
10. Cvekl A, Ashery-Padan R (2014) The cellular and molecular mechanisms of vertebrate lens development. Development 141:4432. https://doi.org/10.1242/dev.107953
11. Danno H, Michiue T, Hitachi K, Yukita A, Ishiura S, Asashima M (2008) Molecular links among the causative genes for ocular malformation: Otx2 and Sox2 coregulate Rax expression. Proc Natl Acad Sci 105:5408–5413. https://doi.org/10.1073/pnas.0710954105
12. Shimozaki K, Zhang CL, Suh H, Denli AM, Evans RM, Gage FH (2012) SRY-box-containing gene 2 regulation of nuclear receptor tailless (Tlx) transcription in adult neural stem cells. J Biol Chem 287:5969–5978. https://doi.org/10.1074/jbc.M111.290403
13. Lodato MA, Ng CW, Wamstad JA, Cheng AW, Thai KK, Fraenkel E, Jaenisch R, Boyer LA (2013) SOX2 co-occupies distal enhancer elements with distinct POU factors in ESCs and NPCs to specify cell state. PLoS Genet 9:e1003288. https://doi.org/10.1371/journal.pgen.1003288
14. Yamamoto S, De D, Hidaka K, Kim KK, Endo M, Sugiyama H (2014) Single molecule visualization and characterization of Sox2-Pax6 complex formation on a regulatory DNA element using a DNA origami frame. Nano Lett 14:2286–2292. https://doi.org/10.1021/nl4044949

15. Sontz PA, Muren NB, Barton JK (2012) DNA charge transport for sensing and signalling. Acc Chem Res 45:1792–1800. https://doi.org/10.1021/ar3001298
16. Andrew WA, Michael A, Stephen LM, Robert JC, Harry BG (1988) Distance dependence of photoinduced long-range electron transfer in zinc/ruthenium-modified myoglobins. J Am Chem Soc 110:435–439. https://doi.org/10.1021/ja00210a020
17. DeRosa MC, Sancar A, Barton JK (2005) Electrically monitoring DNA repair by photolyase. Proc Natl Acad Sci USA 102:10788–10792. https://doi.org/10.1073/pnas.0503527102
18. Boon EM, Livingston AL, Chimiel NH, David SS, Barton JK (2003) DNA-mediated charge transport for DNA repair. Proc Natl Acad Sci USA100, 12543–12547. https://doi.org/10.1073/pnas.2035257100
19. Yavin E, Boal AK, Stemp ED, Boon EM, Livingston AL, O'Shea VL, David SS, Barton JK (2005) Protein-DNA charge transport: redox activation of a DNA repair protein by guanine radical. Proc Natl Acad Sci USA 102:3546–3551. https://doi.org/10.1073/pnas.0409410102
20. Kim S, Li Y, Sancar A (1992) The third chromophore of DNA photolyase: Trp-277 of Escherichia coli DNA photolyase repairs thymine dimers by direct electron transfer. Proc Natl Acad Sci USA 89:900–904
21. Sancar A (2003) Structure and function of DNA photolyase and cryptochrome blue-light photoreceptors. Chem Rev 103:2203–2237. https://doi.org/10.1021/cr0204348
22. BehmoarasT Toulme JJ, Hélène C (1981) A tryptophan-containing peptide recognizes and cleaves DNA at apurinic sites. Nature 292:858–859. https://doi.org/10.1038/292858a0
23. Wagenknecht HA, Stemp EDA, Barton JK (2000) Evidence of electron transfer from peptides to DNA: oxidation of DNA-bound tryptophan using the flash-quench technique. J Am Chem Soc 122:1. https://doi.org/10.1021/ja991855i
24. Wagenknecht HA, Rajski SR, Pascaly M, Stemp ED, Barton JK (2001) Direct observation of radical intermediates in protein-dependent DNA charge transport. J Am Chem Soc 123:4400. https://doi.org/10.1021/ja003986l
25. Mayer-Enthart E, Kaden P, Wagenknecht HA (2005) Electron transfer chemistry between DNA and DNA-binding tripeptides. Biochemistry 44:1749–1757. https://doi.org/10.1021/bi0504557
26. Tashiro R, Wang AH, Sugiyama H (2006) Photoreactivation of DNA by an archaeal nucleoprotein Sso7d. Proc Natl Acad Sci USA 103:16655–16659. https://doi.org/10.1073/pnas.0603484103
27. Sugiyama H, Tsutsumi Y, Saito I (1990) Highly sequence-selective photoreaction of 5-bromouracil-containing deoxyhexanucleotides. J Am Chem Soc 112:6720–6721. https://doi.org/10.1021/ja00174a046
28. Sugiyama H, Fujimoto K, Saito I (1996) Evidence for intrastrand C2' hydrogen abstraction in photoirradiation of 5-halouracil-containing oligonucleotides by using stereospecifically C2'-deuterated deoxyadenosine. Tetrahedron Lett 37:1805–1808. https://doi.org/10.1016/0040-4039(96)00123-2
29. Sugiyama H, Tsutsumi Y, Fujimoto K, Saito I (1993) Photoinduced deoxyribose C2' oxidation in DNA. Alkali-dependent cleavage of erythrose-containing sites via a retroaldol reaction. J Am Chem Soc 115:4443–4448. https://doi.org/10.1021/ja00064a004
30. Saha A, Hashiya F, Kizaki S, Asamitsu S, Hashiya K, Bando T, Sugiyama H (2015) A novel detection technique of polyamide binding sites by photo-induced electron transfer in ^{Br}U substituted DNA. Chem Commun 51:14485. https://doi.org/10.1039/C5CC05104E
31. Hashiya F, Saha A, Kizaki S, Li Y Sugiyama H (2014) Locating the uracil-5-yl radical formed upon photoirradiation of 5-bromouracil-substituted DNA. Nucleic Acids Res 42:13469–13473. https://doi.org/10.1093/nar/gku1133

32. Lewis FD, Wu Y (2001) Dynamics of superexchange photoinduced electron transfer in duplex DNA. J Photochem Photobiol C Photochem Rev 2:1–16. https://doi.org/10.1016/S1389-5567(01)00008-9
33. Watanabe T, Bando T, Xu Y, Tashiro R Sugiyama H (2005) Efficient generation of 2'-deoxyuridin-5-yl at 5'-(G/C)AA(X)U(X)U-3' (X = Br, I) sequences in duplex DNA under UV irradiation. J Am Chem Soc 127:44–45. https://doi.org/10.1021/ja0454743
34. Watanabe T, Tashiro R, Sugiyama H (2007) Photoreaction at 5'-(G/C)AA(Br)UT-3' sequence in duplex DNA: efficient generation of uracil-5-yl radical by charge transfer. J Am Chem Soc 129:8163–8168. https://doi.org/10.1021/ja0692736
35. Jolma A, Yin Y, Nitta KR, Dave K, Popov A, Taipale M, Enge M, Kivioja T, Morgunova E, Taipale J (2015) DNA-dependent formation of transcription factor pairs alters their binding specificity. Nature 527:384. https://doi.org/10.1038/nature15518
36. Xu Y, Tashiro R, Sugiyama H (2007) Photochemical determination of different DNA structures. Nat Protoc 2:78–87. https://doi.org/10.1038/nprot.2006.467

Chapter 6
UVA Irradiation of ^{Br}U-Substituted DNA in the Presence of Hoechst 33258

Abstract Given that our knowledge of DNA repair is limited because of the complexity of the DNA system, a technique called UVA micro-irradiation has been developed that can be used to visualize the recruitment of DNA repair proteins at double-strand break (DSB) sites. Interestingly, Hoechst 33258 was used under micro-irradiation to sensitize 5-bromouracil (^{Br}U)-labeled DNA, causing efficient DSBs. However, the molecular basis of DSB formation under UVA micro-irradiation remains unknown. Herein, we investigated the mechanism of DSB formation under UVA micro-irradiation conditions. Our results suggest that the generation of a uracil-5-yl radical through electron transfer from Hoechst 33258 to ^{Br}U caused DNA cleavage preferentially at self-complementary 5′-AA^{Br}U^{Br}U-3′ sequences to induce DSB. We also investigated the DNA cleavage in the context of the nucleosome to gain a better understanding of UVA micro-irradiation in a cell-like model. We found that DNA cleavage occurred in both core and linker DNA regions although its efficiency reduced in core DNA.

Keywords Micro-irradiation · Electron transfer · 5-Bromouracil
DNA photoreaction · Nucleosome

6.1 Introduction

Ionizing radiation causes double-strand breaks (DSBs) in DNA, leading to severe damage to chromosomes in cells. Nuclear proteins accumulate at DNA-damaged sites to execute DNA repairing functions [1–9]. In the field of molecular cell biology, the behavior of DNA repair proteins induced by DSB formation has been studied extensively. However, our understanding of the repair mechanism of DNA remains limited because of the complexity of the DNA repair process; many of the proteins involved in DNA repair remain unidentified. Recently, a sophisticated technique called UVA micro-irradiation was developed to provoke DSBs in living

A. Saha, *Molecular Recognition of DNA Double Helix*, Springer Theses,
https://doi.org/10.1007/978-981-10-8746-2_6

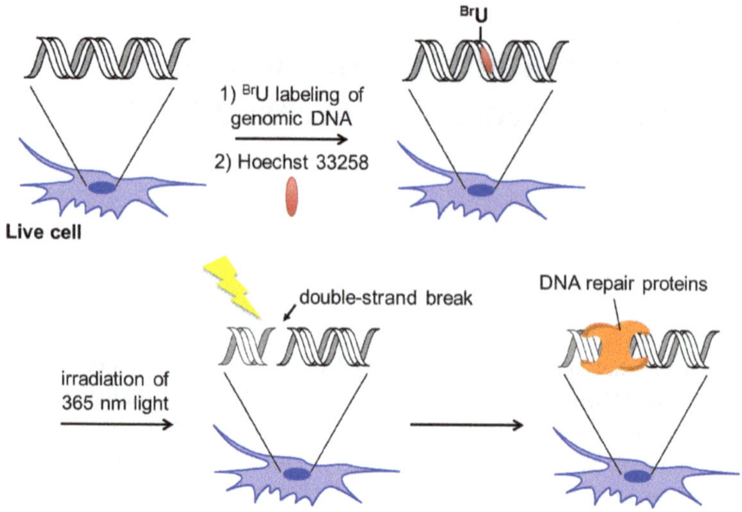

Fig. 6.1 Schematic representation of the UVA micro-irradiation technique, which requires labeling of genomic DNA with d^{Br}UTP followed by sensitization of ^{Br}U-labeled DNA with Hoechst 33258

cells (Fig. 6.1). In this approach, nuclear DNA is labeled with the thymidine analog 5-bromodeoxyuridine (^{Br}U), and the DNA is stained with Hoechst 33258; after UVA laser (365 nm) exposure, single-strand breaks and DSBs are produced [10–12]. By using this technique, it was revealed that DNA repair proteins, Rad51 and hMre11–hRad50 complex, assemble in discrete nuclear loci as part of the cellular response to DNA damage [13]. These studies established the potential of this technique in the field of experimental cell research and radiation biology and led to a better understanding of the DNA repair mechanism.

In this study, we investigated the photoirradiation of ^{Br}U-substituted DNA fragments in the presence of Hoechst 33258. We also investigated the photoirradiation of ^{Br}U-substituted DNA fragments in the context of the nucleosome to gain a better understanding of nucleosome to gain a better understanding of UVA micro-irradiation in a cell-like model.

6.2 Results and Discussions

6.2.1 *Photoreaction on ^{Br}U-Labeled DNA*

To gain insight into Hoechst 33258-induced DNA damage under UVA micro-irradiation, we constructed the DNA1 (298 bp) fragment, in which all the

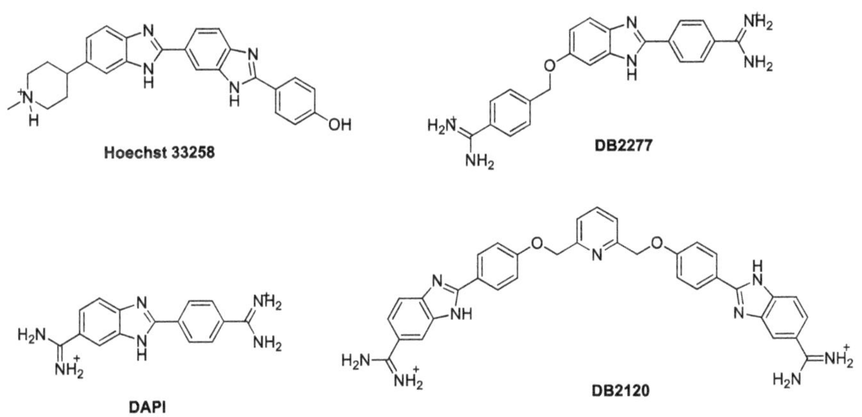

Chart 6.1 Chemical compounds known as DNA minor groove binder used in this study

thymine residues were replaced with ^{Br}U, by performing PCR using d^{Br}UTP instead of dTTP. DNA1 was then irradiated with 365 nm UVA light in the presence of minor-groove-binding Hoechst 33258 [14, 15] and 500 mM isopropanol (Chart 6.1) for 0–4 s. Under these conditions, the presumable uracil-5-yl radical generated upon capturing an electron from the dye abstracts hydrogen from the isopropanol to provide the uracil residue almost quantitatively [16, 17]. The sites containing the uracil residues were then cleaved by uracil glycosylase and heat treatment, and the cleavage sites were subsequently identified by slab gel sequencing. Isopropanol or THF acts as an H-atom donor under in vitro condition, which can trap the reactive radical species [16]. It is also known that without the use of H-atom donor (THF or isopropanol), uracil-5-yl radical can cleave the DNA by abstracting an H-atom from the C1′ position of nearest sugar moiety to produce easily cleavable 2-deoxyribonolactone. Another, simultaneous H-atom abstraction is also known from C2′ α position of the sugar to produce alkali-labile erythrose containing site [18]. As a result, the total DNA cleavage under this condition is lower. But practically, in vivo system of Hoechst-induced DNA damage might consider the generation of labile 2-deoxyribonolactone instead of uracil residues prior to DNA damage. The use of isopropanol followed by UDG enzymatic digestion in the present system is only to produce one photoproduct thus making the DNA damage more visible in the in vitro system. We analyzed DNA cleavage sites for both top and bottom strands of DNA1 by using two differently labeled DNAs in which either the top or the bottom strand of DNA1 was labeled with Texas-Red. As a result, several distinct DNA cleavage bands were observed on the gel (Fig. 6.2a). Importantly, DNA cleavage did not occur in the absence of Hoechst 33258, suggesting that photoinduced electron transfer from Hoechst 33258 to ^{Br}U residues is indispensable for uracil-2-yl radical formation, as reported previously [18, 19].

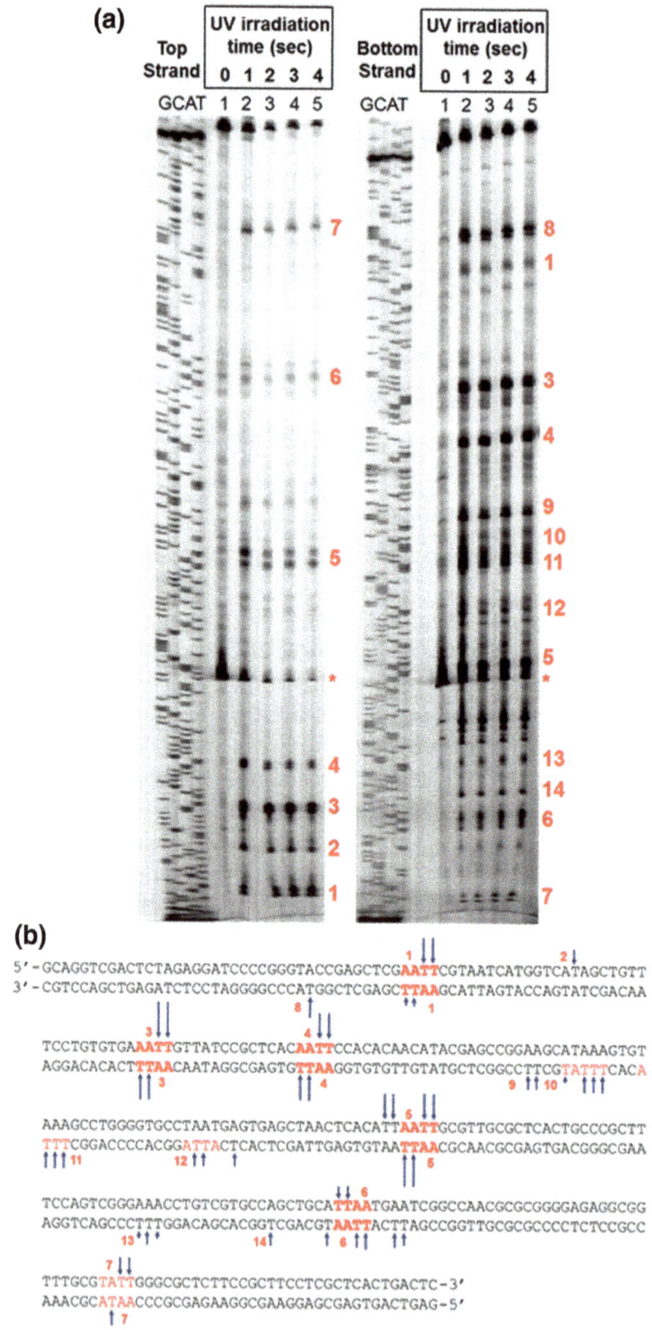

Fig. 6.2 **a** Analysis of DNA1 fragment by using 6% denaturing gel electrophoresis. The DNA cleavage sites are marked from 1 to 14 in both strands. The band marked * derives from secondary structure of the DNA. **b** Mapping of DNA cleavage sites after photoreaction of DNA1 fragment (all Ts were replaced with ^{Br}Us). The cleavage sites are indicated by numbers (from 1 to 14)

The mapping of DNA cleavage sites on DNA1 is shown in Fig. 6.2b. We detected DNA cleavage at Sites 1, 3, 4, 5, 6, and 7. Among them, Sites 1, 3, 4, and 5 correspond to self-complementary 5′-AA^{Br}U^{Br}U-3′ sequences with high intensity. Several DNA cleavages were also observed in other AT-rich sequences, such as Sites 10, 11, and 12. In addition to AT-rich sequences, other DNA cleavage sites (Sites 2, 8, 9, 13, and 14) contained a mix of GC- and AT-rich sequences. It was reported that Hoechst 33258 preferentially binds to AT-rich sequences (especially at 5′-AATT-3′ sequence) [15]. Therefore, this result is consistent with the conclusion that electron transfer occurred from bound Hoechst 33258 to ^{Br}U in the 5′-AA^{Br}U^{Br}U-3′ sequence to produce uracil-5-yl radical in this sequence. In cell, it is thought that the formed uracil-5-yl radical abstracts C1′ hydrogen of deoxyribose from the neighboring 5′-side nucleotide to produce 2′-deoxyribonolactone, which can cause DNA cleavage by following β- and δ-eliminations [20]. Given that 5′-AA^{Br}U^{Br}U-3′ is a self-complementary sequence, it is thought that DNA cleavages at both strands lead to DSB formation.

To examine whether other DNA-binding ligands can cause DNA cleavage, we irradiated ^{Br}U-substituted DNA with other fluorescent chromosomal-staining agents: DAPI, DB2277, and DB2120 (Chart 6.1) with 365 nm irradiation. DAPI is a DNA minor-groove-binding ligand that is normally used for fluorescence imaging. This dye also preferentially recognizes the 5′-AATT-3′ sequence [16]. DB2277 recognizes a single G in sequences such as 5′-AAGTT-3′ through an aza-benzimidazole group and DB2120, a bis-benzimidazole can recognize either A4GT4 or A4T4 sequences [21, 22]. Interestingly, although the three compounds have similar absorption maxima to that of Hoechst 33258 (ca. 350 nm), slab gel sequencing results showed that these compounds failed to cause any strand cleavage in ^{Br}U-substituted DNA (Figs. 6.3 and 6.4). It is thought that compared with these three compounds, Hoechst 33258 more easily donates electron into ^{Br}U by photoexcitation.

6.2.2 *Photoreaction in Nucleosome Structure*

We wondered whether such photoinduced DNA cleavage happens on a nucleosome structure, given that DNA inside a eukaryotic cell is intimately associated with proteins to form the nucleosome; that is, the DNA is wrapped around a histone octamer, which is composed of pairs of H2A, H2B, H3, and H4 [23, 24]. Nucleosomes are assembled into higher order structures, so that genomic DNA can fit into the nucleus. Therefore, we reconstituted a mononucleosome by using ^{Br}U-substituted Widom 601-nucleosome-positioning-sequence-containing 382 bp DNA (DNA2) (Fig. 6.7) [24].

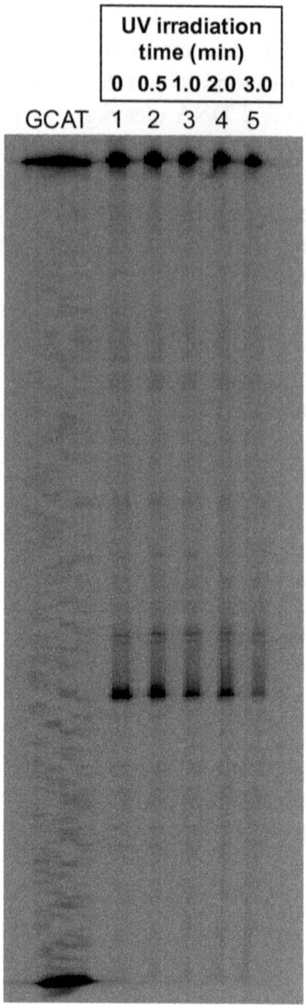

Fig. 6.3 Photoreaction using DAPI on DNA2. The reaction mixture contains 10 nM DNA and 1 μM of DAPI. Irradiation was performed with 365 nm UV for 0–3 min

After preparation of the reconstituted nucleosome, the ^{Br}U-substituted nucleosome was irradiated with Hoechst 33258 and the sites of formation of the uracil-5-yl radical were detected as described previously. At Site 3 (5′-AA$^{Br}U^{Br}U$-3′ sequence), DNA cleavage was significantly reduced on core DNA (Fig. 6.5). Although it is reported that Hoechst 33258 can bind to nucleosomal DNA which face both toward and away from the histone core without affecting the nucleosome structure [25], our result suggests that the binding of Hoechst 33258 to core DNA is

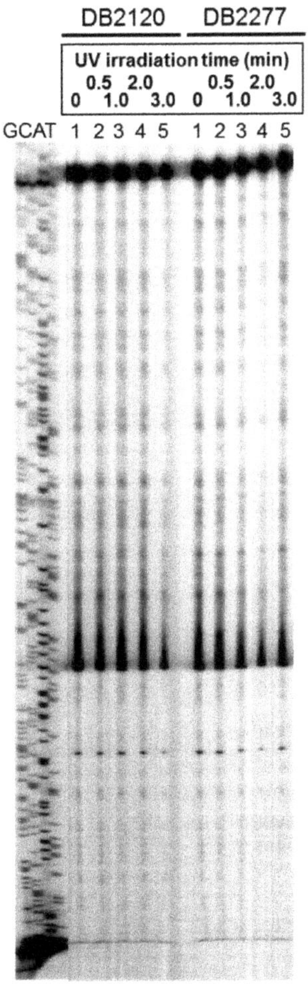

Fig. 6.4 Photoreaction using DB2120 and DB2277 on DNA2 (bottom strand). The reaction mixture contains 10 nM DNA and 1 μM of DB2120 or DB2277. Irradiation was performed with 365 nm UV for 0–3 min

weaker than that of Hoechst 33258 to linker DNA. At Site 2 and Site 4, the efficiency of DNA cleavage was almost same between free DNA and nucleosomal DNA. Its reason seems to be that at both Site 2 and Site 4, the nucleosomal DNA highly fluctuates to enable readily access of Hoechst 33258 to those sites. Previous reports also suggest undisturbed access of small compound and protein to the edges of core DNA [26, 27].

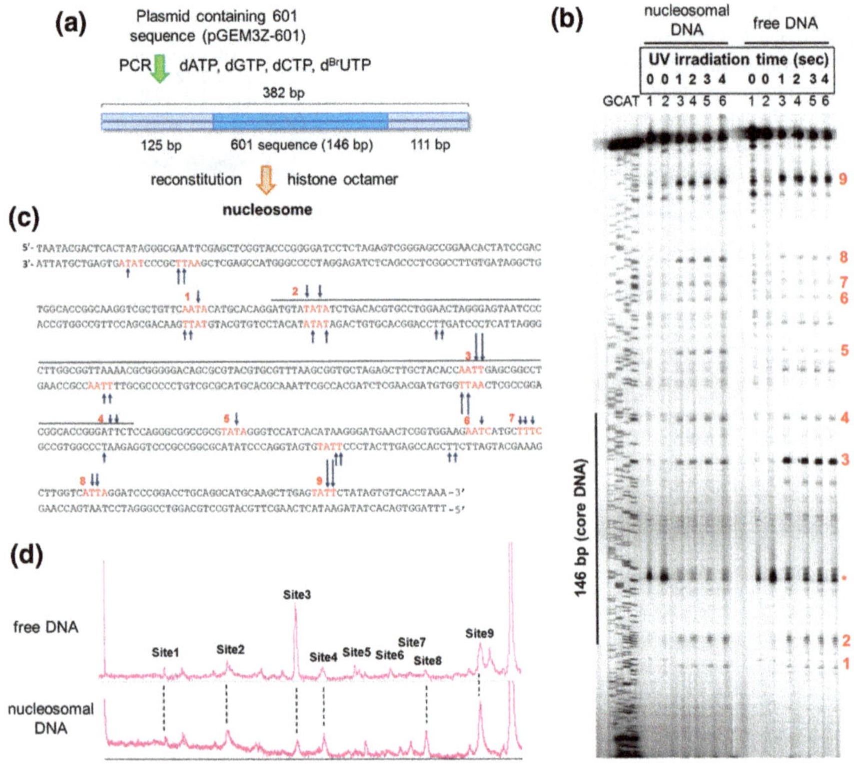

Fig. 6.5 **a** Reconstitution of mononucleosome using 382 bp DNA containing 146 bp of Widom 601 sequence and histone octamer. **b** Analysis of DNA2 fragment by using 6% denaturing gel electrophoresis. The DNA cleavage sites are marked from 1 to 9. The band marked * derives from secondary structure of the DNA. (Lane 1) DNA only (Lanes 2–6) 365 nm irradiation for 0, 1, 2, 3, 4 s with 10 nM DNA and 30 nM Hoechst 33258. Photoreaction result on the nucleosome is shown on the left side, and the photoreaction result in free DNA is shown on the right side. **c** Mapping of DNA cleavage sites after photoreaction of DNA2 top strand (all Ts were replaced with ^{Br}Us). The cleavage sites are indicated by numbers (from 1 to 9). The underlined area in the DNA sequence indicates the 146 bp of core DNA. **d** Densitometric analysis of the cleavage bands in free DNA (top) and in nucleosomal DNA (bottom)

6.3 Conclusion

In summary, we have demonstrated that DNA cleavages occur on ^{Br}U-substituted DNA sensitized with Hoechst 33258 under UVA irradiation. DNA cleavage also occurred in nucleosomal DNA. These results suggest the following mechanism for DSB formation in UVA micro-irradiation: electron transfer from Hoechst 33258 to ^{Br}U occurs to yield a uracil-5-yl radical (Fig. 6.6). In the absence of a hydrogen donor, the uracil-5-yl radical abstracts C1′ hydrogen of the sugar moiety, this can cause single-strand breaks and DSBs in the DNA.

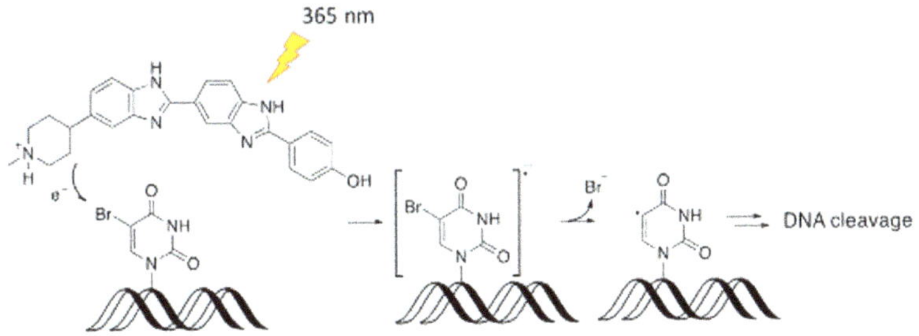

Fig. 6.6 The suggested mechanism of DSB formation by UVA irradiation in the presence of Hoechst 33258

6.4 Materials and Methods

6.4.1 General

Hoechst 33258 and DAPI were purchased from Sigma Aldrich (St. Louis, MO, USA) DB2120 and DB2277 were kindly provided by Prof. David Boykin and Prof. W. David Wilson (Georgia State University, USA).

6.4.2 Preparation of ^{Br}U-Labeled DNA

In this study, we have used two DNA fragments: 298 bp DNA1 and 382 bp DNA2. These two DNAs were amplified by PCR. pUC18 plasmid and pGEM-3z/601 plasmid were used as PCR templates for DNA1 and DNA2, respectively. Primers used for DNA1 amplification; forward primer: 5′-dGCAGGTCGACTCT AGAGGAT-3′, reverse primer: 5′-dGAGTCAGTGAGCGAGGAAG-3′. Primers used for DNA2 amplification; forward primer: 5′-dTAATACGACTCACT ATAGGG-3′, reverse primer: 5′-dATTTAGGTGACACTATAG-3′. For the analysis of top strand, 5′-Texas-Red-labeled forward primer was used, while for the analysis of bottom strand, 5′-Texas-Red-labeled reverse primer was used. All primers were purchased from Sigma Aldrich.

6.4.3 Polymerase Chain Reaction

Master mix for PCR reaction contains: 20 μL of 10 × buffer (500 mM KCl, 100 mM Tris-HCl (pH 8.3), 25 mM $MgCl_2$), 20 μL of each 2 mM dATP, dGTP, dCTP, and d^{Br}UTP, 6 μL of each 10 μM forward and reverse primers, 2 μL of 10 U Taq DNA polymerase, 50 ng of DNA template, and Milli-Q water to total volume of 200 μL. PCR was performed with iCycler (BioRad, Hercules, CA, US) in the following condition: 95 °C for 2 min; 30 cycles of (a) 95 °C for 20 s, (b) 52 °C for 30 s, (c) 68 °C for 30 s; finally 68 °C for 7 min. PCR products were purified with GenElute™ PCR Clean-Up Kit (Sigma Aldrich) and confirmed by 2% agarose gel electrophoresis, quantified by Nano Drop 1000 (Thermo Fisher Scientific, Waltham, MA, USA).

6.4.4 Nucleosome Reconstitution Using 601 Sequence

Texas-Red-labeled DNA (200 nM) and recombinant human histone octamer (EpiCypher, Davis Dr, Durham, NC, USA) (300 nM) were mixed in 2 M NaCl and 20 mM HEPES KOH (pH 7.5) (total volume 50 μL), and placed in Oscillatory Cup (MWCO: 8000) (COSMO BIO, Tokyo, Japan). The dialysis tube was immersed into 500 mL of 2 M NaCl and 20 mM HEPES KOH (pH 7.5) for 2 h at 4 °C, followed by 1.5 M NaCl (overnight), 1.0 M NaCl (8 h), 0.75 M NaCl (overnight), and 0 M NaCl (8 h) (each contains 20 mM HEPES KOH (pH 7.5)). After dialysis, the sample was collected from the tube and stored at 4 °C until use. The formation of Nucleosome was confirmed by native gel eletrophoresis as shown in Fig. 6.7.

6.4.5 Irradiation Set up

LED light (model ZUVC30H, OMRON, Kyoto, Japan) with 300 mW at 365 nm was used for irradiation. The irradiation was performed at 0 °C by keeping the

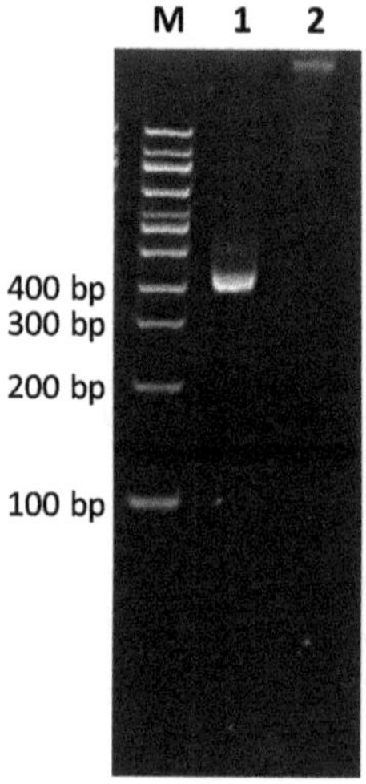

Fig. 6.7 6% native polyacrylamide gel electrophoresis result of free DNA (lane 1) and mononucleosome (lane 2), which was reconstituted using DNA2 (containing 146 bp of Widom 601 sequence) and histone octamer

sample (13 μL) in the cap of 1.5 mL Eppendorf Tube which was placed on a metal plate cooled with ice.

6.4.6 Photoreaction

Photoreaction was performed in 10 nM ^{Br}U-substituted DNA, 10 mM sodium cacodylate (pH 7.0), 500 mM isopropanol, and 30 nm Hoechst 33258 (total volume: 13 μL). After irradiation, the reaction mixture was treated with 1.25 unit of UDG (Takara, Kusatsu, Japan) and incubated at 37 °C for 1 h. The reaction mixture was dried up and 6 μL of gel loading dye was added, followed by heat treatment at 95 °C for 10 min. Then, all of the sample was used for slab gel sequencing.

References

1. Berns MW (1978) The laser microbeam as a probe for chromatin structure and function. Methods Cell Biol 18:277–294
2. Cremer C, Cremer T (1986) Induction of chromosome shattering by ultraviolet light and caffeine: the influence of different distributions of photolesions. Mutat Res 163:33–40
3. Bonner WM, Redon CE, Dickey JS, Nakamura AJ, Sedelnikova OA, Solier S, Pommier Y (2008) GammaH2AX and cancer. Nat Rev Cancer 8:957–967. https://doi.org/10.1038/nrc2523
4. Kinner A, Wu W, Staudt C, Iliakis G (2008) Gamma-H2AX in recognition and signaling of DNA double-strand breaks in the context of chromatin. Nucleic Acids Res 36:5678–5694
5. FitzGerald JE, Grenon M, Lowndes NF (2009) 53BP1: function and mechanisms of focal recruitment. Biochem Soc Trans 37:897–904. https://doi.org/10.1042/BST0370897
6. Misteli T, Soutoglou E (2009) The emerging role of nuclear architecture in DNA repair and genome maintenance. Nat Rev Mol Cell Biol 10:243–254. https://doi.org/10.1038/nrm2651
7. Stucki M, Jackson SP (2006) gammaH2AX and MDC1: anchoring the DNA-damage-response machinery to broken chromosomes. DNA Repair 5:534–543. https://doi.org/10.1016/j.dnarep.2006.01.012
8. Benjdia A, Heil K, Barends TR, Carell T, Schlichting I (2012) Structural insights into recognition and repair of UV-DNA damage by spore photoproduct lyase, a radical SAM enzyme. Nucleic Acids Res 40:9308–9318. https://doi.org/10.1093/nar/gks603
9. Fei J, Kaczmarek N, Luch A, Glas A, Carell T, Naegeli H (2011) Regulation of nucleotide excision repair by UV-DDB: prioritization of damage recognition to internucleosomal DNA. PLoS Biol 9:e1001183. https://doi.org/10.1371/journal.pbio.1001183
10. Limoli CL, Ward JF (1993) A new method for introducing double-strand breaks into cellular DNA. Radiat Res 134:160–169. https://doi.org/10.2307/3578455
11. Suzuki K, Yamauchi M, Oka Y, Suzuki M, Yamashita S (2011) Creating localized DNA double-strand breaks with microirradiation. Nat Protoc 6:134–139. https://doi.org/10.1038/nprot.2010.183

12. Walter J, Cremer T, Miyagawa K, Tashiro S (2003) A new system for laser-UVA-microirradiation of living cells. J Microsc 209:71–75
13. Beishline K, Kelly CM, Olofsson BA, Koduri S, Emrich J, Greenberg RA, Azizkhan-Clifford J (2012) Sp1 facilitates DNA double-strand break repair through a nontranscriptional mechanism. J Mol Cell Biol 32:3790–3799. https://doi.org/10.1128/MCB.00049-12
14. Harshman KD, Dervan PB (1985) Molecular recognition of B-DNA by Hoechst 33258. Nucleic Acids Res 13:4825–4835
15. Breusegem SY, Clegg RM, Loontiens FG (2002) Base-sequence specificity of Hoechst 33258 and DAPI binding to five (A/T)4 DNA sites with kinetic evidence for more than one high-affinity Hoechst 33258-AATT complex. J Mol Biol 315:1049–1061. https://doi.org/10.1006/jmbi.2001.5301
16. Hashiya F, Saha A, Kizaki S, Li Y, Sugiyama H (2014) Locating the uracil-5-yl radical formed upon photoirradiation of 5-bromouracil-substituted DNA. Nucleic Acids Res 42:13469–13473. https://doi.org/10.1093/nar/gku1133
17. Saha A, Kizaki S, De D, Endo M, Kim KK, Sugiyama H (2016) Examining cooperative binding of Sox2 on DC5 regulatory element upon complex formation with Pax6 through excess electron transfer assay. Nucleic Acids Res 44:e125. https://doi.org/10.1093/nar/gkw478
18. Sugiyama H, Fujimoto K, Saito I (1996) Evidence for intrastrand C2' hydrogen abstraction in photoirradiation of 5-halouracil-containing oligonucleotides by using stereospecifically C2'-deuterated deoxyadenosine. Tetrahedron Lett 37:1805–1808. https://doi.org/10.1016/0040-4039(96)00123-2
19. Morinaga H, Takenaka T, Hashiya F, Kizaki S, Hashiya K, Bando T, Sugiyama H (2013) Sequence-specific electron injection into DNA from an intermolecular electron donor. Nucleic Acids Res 41:4724–4728. https://doi.org/10.1093/nar/gkt123
20. Sugiyama H, Tsutsumi Y, Fujimoto K, Saito I (1993) Photoinduced deoxyribose C2' oxidation in DNA. Alkali-dependent cleavage of erythrose-containing sites via a retroaldol reaction. J Am Chem Soc 115:4443–4448. https://doi.org/10.1021/ja00064a004
21. Paul A, Nanjunda R, Kumar A, Laughlin S, Nhili R, Depauw S, Deuser SS, Chai Y, Chaudhary AS, David-Cordonnier MH, Boykin DW, Wilson WD (2015) Mixed up minor groove binders: convincing A·T specific compounds to recognize a G·C base pair. Bioorg Med Chem Lett 25:4927–4932. https://doi.org/10.1016/j.bmcl.2015.05.005
22. Harika NK, Paul A, Stroeva E, Chai Y, Boykin DW, Germann MW, Wilson WD (2016) Imino proton NMR guides the reprogramming of A T specific minor groove binders for mixed base pair recognition. Nucleic Acids Res 44:4519–4527. https://doi.org/10.1093/nar/gkw353
23. Kornberg RD, Lorch Y (1999) Twenty-five years of the nucleosome, fundamental particle of the eukaryote chromosome. Cell 98:285–294
24. Vasudevan D, Chua EY, Davey CA (2010) Crystal structures of nucleosome core particles containing the '601' strong positioning sequence. J Mol Biol 403:1–10. https://doi.org/10.1016/j.jmb.2010.08.039
25. Leslie KD, Fox KR (2002) Interaction of Hoechst 33258 and echinomycin with nucleosomal DNA fragments containing isolated ligand binding sites. Biochemistry 41:3484–3497
26. Zou T, Kizaki S, Pandian GN, Sugiyama H (2016) Nucleosome assembly alters the accessibility of the antitumor agent duocarmycin B2 to duplex DNA. Chem Eur J 22:8756–8758. https://doi.org/10.1002/chem.201600950
27. Kizaki S, Zou T, Li Y, Han YW, Suzuki Y, Harada Y, Sugiyama H (2016) Preferential 5-methylcytosine oxidation in the linker region of reconstituted positioned nucleosomes by tet1 protein. Chem Eur J 22:16598–16601. https://doi.org/10.1002/chem.201602435

Curriculum Vitae

Dr. Abhijit Saha

Department of Chemistry, Graduate School of Science, Kyoto University, Kyoto, Japan

Current Address
Institut Curie/Centre de recherche
UMR9187/U1196
Batiment 110, Chemistry, Modeling and Imaging for biology
15 Rue Georges Clémenceau
91405 Orsay, France
Tel: +33-7.83.77.96.64
Email: abhijit.saha@curie.fr

Education

- Ph.D. in Chemical Biology, Graduate School of Science, Kyoto University, Japan (April 2012–March 2015).
  Supervisor: **Prof Hiroshi Sugiyama**, Chemical Biology Laboratory (Sugiyama Lab).
  Dissertation Title: Chemical Biology Approaches for the molecular recognition of DNA double helix.

A. Saha, *Molecular Recognition of DNA Double Helix*, Springer Theses,
https://doi.org/10.1007/978-981-10-8746-2

- M.Sc. in Organic Chemistry, Cotton College, Gauhati University, India (August 2006–August 2008).
- B.Sc. in Chemistry (Honors) with Physics, Mathematics, and English.
  Madhab Chaudhury College, Gauhati University, India (June 2002–June 2006).

Current Position

- Post Doctoral Research Fellow at Institut Curie, France.
  Supervisor: **Dr. Marie-Paule Teulade-Fichou**, (UMR9187-U1196, Institut Curie, Campus Universitaire, Bat 110, 91405 Orsay, France).
  Research Interest: Chemical Biology of G-Quadruplex DNA.

Awards

- Received Doctoral Fellowship from Seiwa International Scholarship Foundation April 2013.
- Received iCeMS, Kyoto University overseas travel grant for visiting John Hopkins University, Boston University, and Gordon Research Conference in June 2014.
- Received outstanding poster award at Gordon Research Conferences on Bioorganic Chemistry 2014 held at Proctor Academy, New Hampshire (June 8–13). (Title: Synthesis and Biological Evaluation of Targeted Transcriptional Activator that Could Switch ON the Multiple Pluripotency Genes in Mouse Fibroblast).
- Qualified **GATE** (Graduate Aptitude Test in Engineering) [All India Rank 370].
- Received Junior Research Fellowship from Council of Scientific and Industrial Research (**CSIR**), NET-LS [All India Rank 26].

Professional Experience

- Worked as Junior Research Fellow at the **Indian Institute of Technology (IIT) Guwahati** from August **2009 to April 2012** on a project entitled "β-breaker dipeptide: development of a novel approach for amyloid disruption" under the supervision of Dr. Bhubaneswar Mandal (Associate Professor, Indian Institute of Technology Guwahati, India).

Lightning Source UK Ltd.
Milton Keynes UK
UKHW02n1416190418
321331UK00001B/52/P

9 789811 087455